THE NEW SCIENCE ENCYCLOPEDIA

Tom Jackson and
Janet Bingham

ARCTURUS

Picture Credits

Every attempt has been made to clear copyright. Should there be any inadvertent omission, please apply to the publisher for rectification.

Alamy Stock Photo: 14cl PA Images, 14tl James Arnold, 20–21c Rings and jewelery, 24–25c imageBROKER, 26c Robert Brook/Science Photo Library, 28–29c Image Source, 32c Dorling Kindersley Ltd, 32tl Dorling Kindersley Ltd, 34–35c H. Mark Weidman Photography, 44–45c RooM the Agency, 45br Historic Images, 45c gloriasalgado, 73br Archive PL, 73tl Science History Images, 74–75 Nick Upton, 84cr Paul Ridsdale, 86br Morphart Creation 88–89c Simon Balson, 90cl Oliver Furrer, 96–97c Jake Lyell, Inc., 103tr Colport, 104bl Pictorial Press Ltd, 110cl ClassicStock, 114–115c US Navy Photo, 118–119c Paul Hennessy, 118cr Colin Underhill, 120–121c Dennis Cox, 127tl Conspectus, 134bc David R. Frazier Photolibrary, 139br IanDagnall Computing, 142–143c sciencephotos, 148cl Ferenc Szelepcsenyi, 151tr Science History Images, 176bl Universal Images Group North America LLC, 184br INTERFOTO, 190–191c Antje Schulte/Aphid World, 220–221c Worldpix, 231tr Chroma Collection, 244cl All Canada Photos, 246bl Doug Perrine, 248–249c BIOSPHOTO, **Getty Images:** 124cl salihkilic, 202cr Corbis/VCG; **Nature Picture Library:** 178–179c, 187tr; **Science Photo Library:** 27c TOM HOLLYMAN, 40–41c MARTYN F. CHILLMAID, 41tr Turtle Rock Scientific, 40br SCIENCE SOURCE, 42–43c Michael Penev/US DEPARTMENT OF ENERGY, 46cl Alexandre Dotta/Science Source, 56–57c SAKKMESTERKE, 94br CERN, 136tr, 182c Steve Gschmeissner, 194br Ken Welsh/Design Pictures, 202cl Kermoal/BSIP; **Shutterstock:** 10–11c New Africa, 10br Lilkin, 10cl Albertm24, 11br fortton, 12–13c kmls, 13c Kostsov, 13tr Prostock-studio, 14b Net Vector, 15tr Hecos, 16–17c VasiliyBudarin, 16bl successo images, 16br petroudny35, 16tr Andy Dean Photography, 18–19c Bedrin, 18bl Atlantist Studio, 18tr anchalee thaweeboon, 19tr zizou7, 20br, 20cl Charles Shapiro, 22–23c Idutko, 22cl peterschreiber.media, 23bc MarcelClemens, 23c Sansanorth, 23tl Kedar Vision, 24bl zizou7, 24tl LoopAll, 26–27c New Africa, 26b Lookiepix, 28c Jo Sam Re, 28tl Jo Sam Re, 29tr IT Tech Science, 30–31c Robert Kneschke, 30c Sergey Merkulov, 31tl ggw, 32–33c William Perugini, 33tr ANATOLY Foto, 34cr Mark Agnor, 35cr Sokor Space, 36–37c Halfpoint, 36br Maria Symchych, 36cl Stephen Mcsweeny, 37br nguyen thi phuong dieu, 37tr Incomible, 38–39c 2DAssets, 38–39c Art Stocker, 39br ElenaTlt, 40bl trgrowth, 40cl 2DAssets, 42bl Evgeniyqw, 42cr Douglas Cliff, 44bl YARUNIV Studio, 44cr 2DAssets, 44cr trgrowth, 45tr Nataliia Budianska, 46–47c Max Topchii, 46bl StudioMolekuul, 47bc Kitch Bain, 47tl New Africa, 48–49c Ivan Kurmyshov, 48c Tawansak, 49bl Feng Yu, 49tl Roman Zaiets, 50–51c Tina Gutierrez, 50tr Jo Sam Re, 52–53 Pozdeyev Vitaly, 52c Sergey Merkulov, 53bc Ake13bk, 54c SritanaN, 54cl Adisak Riwkratok, 55tc Ali DM, 56l buteo, 57bl Egoreichenkov Evgenii, 57cr OSweetNature, 58–59c DeepSkyTX, 58cr OSweetNature, 59br Ad_hominem, 60–61c DoJed YeT, 60bl VectorMine, 60cr Photoongraphy, 62–63c Francesco ConT, 62c VectorMine, 63cr Vladimir Arndt, 64–65c Dragos AsaWei, 64tr Kichigin, 64tr Merkushev Vasiliy, 65b Maleo, 66–67c Vitaliy Karimov, 66bl Peter Gudella, 66cr GraphicsRF.com, 68–69c Melissa Burovac, 68cl Seqoya, 68br TinaSova20, 70bl Dark Moon Pictures, 70cr petrroudny35, 72–73c hutch photography, 72cr DKN0049, 74bl ju_see, 75bl LillieGraphie, 75tr popcorn-arts, 76–77c winnond, 76br Beautiful landscape, 77br Sandy Storm, 77tl ShadeDesign, 80–81c Stock Rocket, 80br Sirocco, 80cl Aygul Sarvarova, 82–83c JoeSAPhotos, 82bl Anatoliy Sadovskiy, 82c Pixel B, 84–85c New Africa, 85bl Roman Samborskyi, 86–87c Sky Antonio, 86cl Kirschner, 88c StockCanarias, 88tc Artsiom P, 90–91c Coolakov_com, 90br Rabilbal poudel, 91tr LeManna, 92 cr Gonzalo Buzonni, 92–93c New Africa, 92bl Stas Moroz, 94–95c solarseven, 94cl petrroudny43, 96br Maridav, 96c hedgehog96, 98bl Tomas Ragina, 98r Angelaoblak, 98cl 2DAssets, 100–101c txking, 100bl JoeSAPhotos, 100cl Rupendra Singh Rawat, 102–103c Lianys, 102br Friends Stock, 102cl Mike Flippo, 104–105c Monkey Business Images, 104bl StoryTime Studio, 105tl PeopleImages.com Yuri A, 106–107c RobSt, 106br P Greenwood Photography, 107cr sirtravelalot, 108–109 Shcherbakov Ilya, 108–109c Koto Images, 108bl Rawpixel.com, 108c BearFotos, 110br Starodubtsev Konstantin, 112–113c Xmentoys, 112br s.dali, 112c BlueRingMedia, 113br Artur Didyk, 114–115c AlexLMX, 114br GLYPHstock, 114cl Nick Fox, 116–117c Wut.Anunai, 116cl Rodrigo Garrido, 117c Benoist, 118bl BrainCityArts, 120cl BiniClick, 121tr Dejan Dundjerski, 122cl Arina P Habich, 122–123c Littlekidmoment, 122bc artjazz, 124–125c Grindstone Media Group, 125br Aerial-motion, 126br GoXxu Chocolate, 127cl George Howard Jr, 128–129c Christian Bertrand, 128c EpicStockMedia, 129br petrroudny43, 130–131c ONYXprj, 130br michaeljung, 130cl Subbotina Anna, 132–133c VeeJay, 132c Nutkins.J, 133bc Natee Meepian, 134–135c Anton Brehov, 134cl tersetki, 135br sree rag, 136–137c Kuki Ladron de Guevara, 136cl VectorMine, 138–139c EternalMoments, 138bl Smeerjewegproducties, 140–141c Ollyy, 138c Hannah Spray Photography, 140br kotikoti, 140cl KRPD, 142br New Africa, 142cl Johan Swanepoel, 144–145c NewSs, 144bl VectorMine, 144br Billion Photos, 146–147c Vershinin91, 146c VectorMine, 147tl petrroudny43, 148–149c Lappo Alexander, 148bl Andrewshots, 150–151c ronstik, 150bl huntingSHARK, 150c huntingSHARK, 152–153c Gerry H, 152c Designua, 153tr Simon_Annable, 154–155c ssuaphotos, 154bl Evgeny_V, 154c zstock, 156–157c Grzegorz Czapski, 156cl Rainbow06, 156cr Dn Br, 157br Peter Sobolev, 160bl Wassana Mathipikhai, 160cr kamomeen, 160–161c Ryszard Filipowicz, 161tl Ajit S N, 161t Crystal Alba, 161t Dennis W Donohue, 161t G-Stock Studio, 162–163c mchenke, 162cl Kletr, 163cr Jacek Chabraszewski, 164–165c Lebendkulturen.de, 164b Dr. Norbert Lange, 164cl 3d_man, 165tr Morphart Creation, 166–167c Ivanova Ksenia, 166cl Hanahstocks, 167tl Kuttelvaserova Stuchelova, 168–169c ON-Photography Germany, 168br Evgeny Atamanenko, 168cl Passakorn Shinark, 169tl chanus, 170–171c blue-sea.cz, 170c Vojce, 171tr Marko Blagoevic, 172–173c Prathankarnpap, 172bl Hugh Lansdown, 172cr mar_chm1982, 173tr Witsawat.S, 174–175c J. Esteban Berrio, 174–175c Norjipin Saidi, 174cl DedeDian, 175tl Tennessee Witney, 176c Epidavros, 176cr moosehenderson, 177br Everett Collection, 178bl Paul Vinten, 179br lev radin, 179tr idizi, 180–181c Golden Family Foto, 180br Ldarin, 180cl Sakurra, 182–183c Mazur Travel, 182bl Choksawatdikorn, 184–185c Midori Photography, 184cl BlueRingMedia, 185cr Designua, 186–187c Kurit afshen, 186bc Dave Hansche, 186cl kooanan007, 188–189c Janelle Lugge, 188c Peter-Hg, 189cr MSMondadori, 190cl StevenWhitcherPhotography, 191br A.Sych, 191tl olga gl, 192–193c Prostock-studio, 192bl Valentina Antuganova, 192tl EreborMountain, 194–195c Mladen Mitrinovic, 194bl Designua, 195c adriaticfoto, 196–197c aslysun, 196–197c Tridsanu Thopet, 196br Neveshkin Nikolay, 196c Biology Education, 196cl Oleksandr Drypsiak, 197tr Phonlamai Photo, 197tl VectorMine, 198br Olga Popova, 198c Alex Mit, 198c Peter Hermes Furian, 199cr DestinaDesign, 200–201c Air Images, 200cl VectorMine, 201cr Jose Luis Calvo, 201t Designua, 202–203c sutadimages, 202br VectorMine, 204–205c Maksim Denisenko, 204bl Designua, 204cr solar22, 205tl VectorMine, 206–207c Dragana Gordic, 206br peterschreiber.media, 206cl cunaplus, 208–209c bbernard, 208br VectorMine, 208cl VectorMine, 209cr German Vizulis, 210–211c aslysun, 210br murat photographer, 210cl VectorMine, 212–213c Yusev, 212bl SP-Photo, 212c Sakurra, 213b Carolina K. Smith MD, 214–215c Monkey Business Images, 214br Focused Adventures, 214cl Sakurra, 215br Choksawatdikorn, 216–217c Drazen Zigic/iStock, 216c White Space Illustrations, 217br OlgaReukova, 217br Alila Medical Media, 218–219c Grossinger, 218b Katakari, 218bl Designua, 220b Manee_Meena, 220cr Maryna Olyak, 221tr StudioMolekuul, 222–223c Iokanan VFX Studios, 222b Designua, 222cl Greentellect Studio, 224–225c View Apart, 224c Emre Terim, 225b Dee-sign, 225tr ShadeDesign, 226 Ody_Stocker, 226–227c Nixx Photography, 226br Dee-sign, 228–229c koya979, 228c Achiichii, 228c VikiVector, 229br GraphicsRF.com, 230–231c Pixel_life, 230br Everett Collection, 230cl Joaquin Corbalan P, 232–233c Ryan Boedi, 232bl khlungcenter, 232cr EcoSpace, 234–235c Tunatura, 234c Triff, 235br Anne Coatesy, 236–237c Visual Storyteller, 236br DiBtv, 237c Iri_sha, 238–239c CoreRock, 238bl Laura Pl, 238c VectorMine, 240–241c Designua, 240cl Fotokostic, 241b Merkushev Vasiliy, 242–243 Ali A Suliman, 242bl vladsilver, 242c G.Lukac, 244–243c Teo Tarras, 245tr Creative Travel Projects, 246–247c NaturePicsFilms, 246c EreborMountain, 248br frank60, 248cl MVolodymyr, 250–251c Sergey Uryadnikov, 250bl Oasishifi, 250c Jacques Sztuke, **University of Illinois** Archives: 54br; **Wellcome Collection:** 29br, 43tr, 51tr, 67tr; **Wikimedia Commons:** 11tr, 12br, 15tr, 19tr, 21tr, 22b MRC Laboratory of Molecular Biology, 24br, 30br, 32br Harvard University, 35br, 39tr, 48br, 52bl, 52tr, 57tr, 59tr Smithsonian Institution, 61br, 62br, 64br, 69br, 69tr, 81tr, 83br, 85br, 87tr, 88br BorderlineRebel, 93br Palace of Versailles, 95br KPNO/NOIRLab/NSF/AURA, 97tr, 98br, 101br, 106br, 109br, 111tr, 115tr, 116br, 119br, 120br MRC Laboratory of Molecular Biology, 122br, 124br, 125br, 128, 131tr, 132br, 134br Michel Bakni, 139tr, 143br, 145tr, 146br, 149br, 153br, 156br, 161bl, 162br Howard University, 166br, 169br, 170br, 174tr, 181br, 183tr, 188br, 193br, 194br, 203br, 205br Michel Bakni, 207br Stefan Albrecht/BioNTech, 211tr, 213tr, 215tr, 216br, 219br, 223tr, 224br MRC Laboratory of Molecular Biology, 226br, 234br, 234br Steve Jurvetson, 237br, 238br, 240br Bruno Comby, 243br Augustus Binu, 244br Kingkongphoto & www.celebrity-photos.com, 247br, 249tr Daniela.cardenas, 250br. Cover images all **Shutterstock:** back cover tl Carolina K. Smith MD, bl Golden Family Foto, c AlexLMX tr LiliGraphie, br EpicStockMedia; spine t Kichigin b VectorMine; front cover balloon artpritsadee, zebra Herticet, atom Sharp Dezign, butterfly Mircea Costina, cell Marco G Faria, mushrooms MrRGraz

ARCTURUS

This edition published in 2024 by Arcturus Publishing Limited
26/27 Bickels Yard, 151–153 Bermondsey Street,
London SE1 3HA

Copyright © Arcturus Holdings Limited

All rights reserved. No part of this publication may be reproduced, stored in a retrieval system, or transmitted, in any form or by any means, electronic, mechanical, photocopying, recording, or otherwise, without prior written permission in accordance with the provisions of the Copyright Act 1956 as amended. Any person or persons who do any unauthorized act in relation to this publication may be liable to criminal prosecution and civil claims for damages.

ISBN: 978-1-3988-4106-2
CH011572US
Supplier 29, Date 0624, PI 00006463
Printed in China

Author of Physics and Biology sections: Tom Jackson
Author of Chemistry section: Janet Bingham
Consultants: Steve Parker, Anne Rooney, and Robert Snedden
Designer: Lorraine Inglis
Picture research: Paul Futcher and Lorraine Inglis
Cover design and illustrations: Paul Oakley
Editor: Donna Gregory

CONTENTS

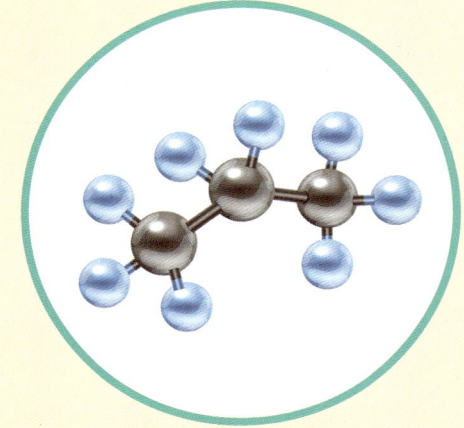

INTRODUCTION 6

CHEMISTRY 8

Chapter 1
Matter, Atoms, and Molecules

States of Matter	10
Particle Theory	12
Changing States	14
Mixtures and Solutions	16
Atomic Structure	18
Elements	20
Molecules	22
Compounds	24
Covalent Bonds	26
Ionic Bonds	28
Reactions	30
Chemical Reactions	32
Exothermic and Endothermic Reactions	34
Important Properties	36

Chapter 2
Elements and the Periodic Table

The Periodic Table	38
Element Groups	40
Tiny Hydrogen	42
Alkaline Earth Metals	44
The Halogens	46
Noble Gases	48
Metals	50
Non-metals and Semi-metals	52
Inorganic and Organic Chemicals	54
Radioactivity	56

Chapter 3
Chemistry in Nature

The First Chemistry	58
Our World	60
Rocks and Minerals	62
Wonderful Water	64
Creative Carbon	66
Essential Oxygen	68
Team Nitrogen	70
Crucial Glucose	72
Plant Chemicals	74
Body Chemistry	76

PHYSICS 78

Chapter 4
Forces and Matter

Matter and Energy	80
Electromagnetism	82
Magnets	84
Gravity	86
Weight and Mass	88
Friction and Drag	90
Pressure	92
Dark Matter	94

Chapter 5
Energy and Motion

Doing Work	96
Thermal Energy	98
Kinetic Energy	100
Potential Energy	102
Other Types of Energy	104
Power	106
Levers	108
Ramps and Screws	110
Wheels and Pulleys	112
Engines	114
First Law of Motion	116
Second Law of Motion	118
Third Law of Motion	120
Momentum	122
Velocity and Acceleration	124

Chapter 6
Waves and Optics

Properties of a Wave	126
Types of Waves	128
Electromagnetic Spectrum	130
Interference	132
Reflection	134
Refraction	136
Lenses	138
Distortion and Diffraction	140

Chapter 7
Electricity

What Is Electricity?	142
Conductors and Insulators	144
Current	146
Voltage	148
Circuits	150
Electric Power	152
Renewable Power	154
Electronics	156

BIOLOGY 158

Chapter 8
Tree of Life

Classification	160
Bacteria	162
Protists	164
Plants	166
Fungi	168
Simple Invertebrates	170
Arthropods	172
Lower Vertebrates	174
Birds	176
Mammals	178

Chapter 9
Plant and Animal Biology

Photosynthesis and Respiration	180
Plant Bodies	182
Plant Reproduction	184
Animal Bodies	186
Animal Locomotion	188
Animal Reproduction	190
Digestion and Excretion	192
Respiratory System	194
Circulatory System	196
Skeleton	198
Muscles	200
Nervous System	202
The Senses	204
Immune System and Diseases	206
Reproductive System	208

Chapter 10
Cells and Genetics

Studying Cells	210
Plant Cell	212
Animal Cell	214
Bacterial Cell	216
Cell Membranes and Transport	218
Enzymes	220
Cell Division	222
DNA and Chromosomes	224
Meiosis	226
Sex Cells and Fertilization	228
The Theory of Evolution	230
Sexual Selection	232

Chapter 11
Habitats and Ecosystems

Ecosystems	234
Food Webs	236
Carbon Cycle	238
Other Cycles	240
Dry Biomes	242
Wet Biomes	244
Ocean Zones	246
Symbiosis	248
Social Groups	250

GLOSSARY 252
INDEX 255

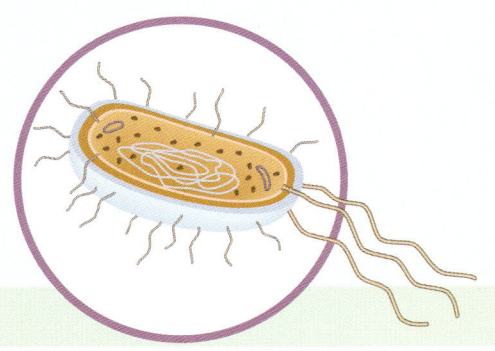

Introduction

We humans are naturally curious creatures. When our ancestors gazed up at the night sky or watched the changing seasons, they would have wondered why the stars shone, or why leaves fell from trees.

Today, we are in the happy position of being able to answer these and many other questions. This book draws together knowledge from different disciplines—chemistry, physics, and biology. Each of these subject areas is like a different lens through which we can see our Universe.

If a scientist doesn't understand something, they ask a question. They then make predictions and form a hypothesis about what the answer might be. A hypothesis is a suggested explanation that gives a starting point to investigate.

Chemistry

Chemistry is the study of the "stuff" the world is made from—matter. All matter is made from tiny building blocks called atoms. Chemistry explains how atoms join together to create every material we know. So chemistry is going on all around us, every day—and lies behind lots of the things we take for granted in our lives, from computers to clothes and food.

Some experiments are done in a controlled space such as a laboratory. Others are done "in the field." For any experiment, a scientist must determine their method—the steps they will take and repeat to ensure the experiment works. Safety measures must be considered as well!

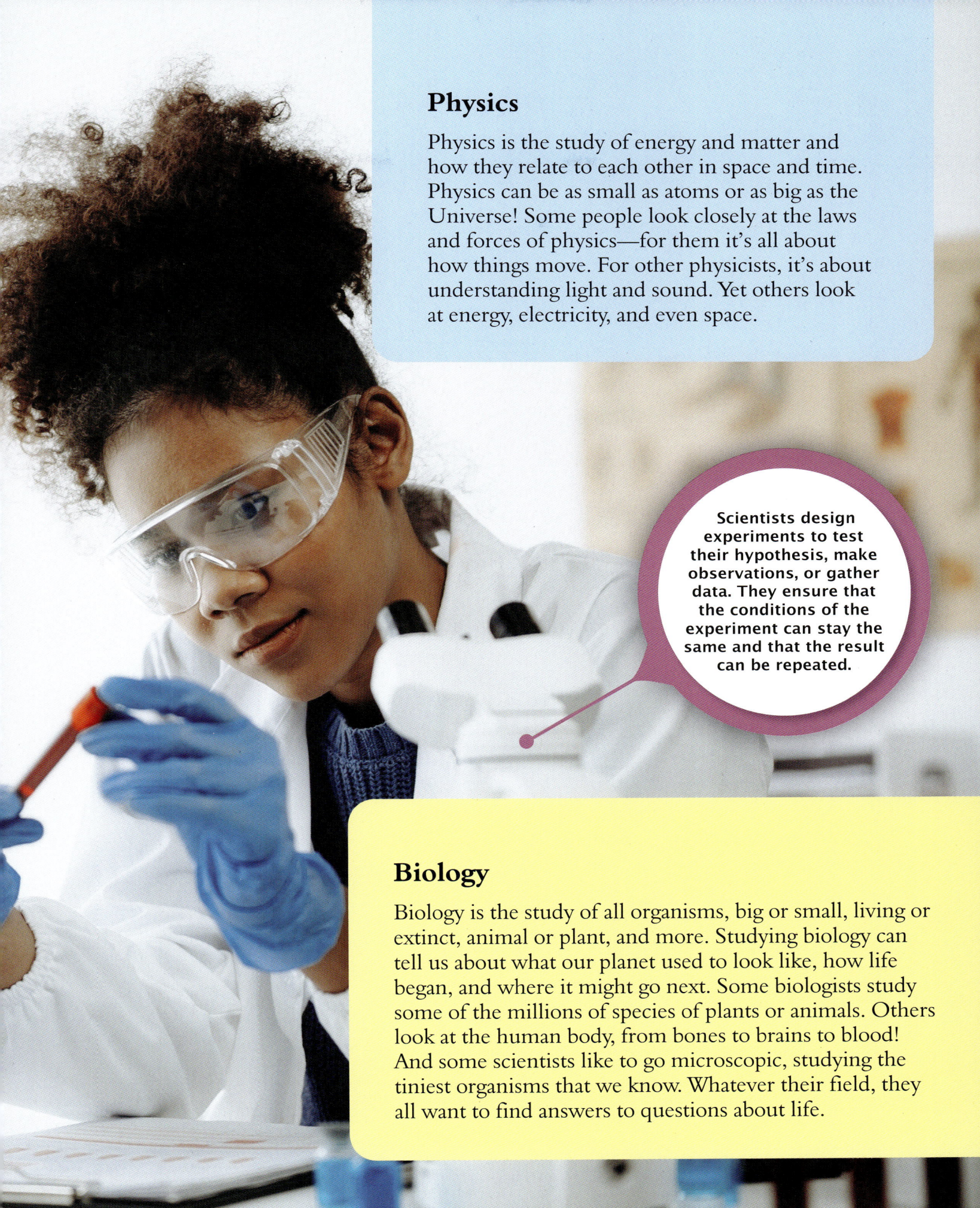

Physics

Physics is the study of energy and matter and how they relate to each other in space and time. Physics can be as small as atoms or as big as the Universe! Some people look closely at the laws and forces of physics—for them it's all about how things move. For other physicists, it's about understanding light and sound. Yet others look at energy, electricity, and even space.

Scientists design experiments to test their hypothesis, make observations, or gather data. They ensure that the conditions of the experiment can stay the same and that the result can be repeated.

Biology

Biology is the study of all organisms, big or small, living or extinct, animal or plant, and more. Studying biology can tell us about what our planet used to look like, how life began, and where it might go next. Some biologists study some of the millions of species of plants or animals. Others look at the human body, from bones to brains to blood! And some scientists like to go microscopic, studying the tiniest organisms that we know. Whatever their field, they all want to find answers to questions about life.

CHEMISTRY

Chemistry is the science of stuff. It investigates everything there is to know about different substances—whether they are natural, like rocks and water, or made by humans, such as plastics. Chemistry also investigates substances in living things, including the human body. No matter where a substance is found, it will always follow scientific rules. There are many millions of substances for chemists to study in different ways, including hundreds that we rely on each day.

Matter, Atoms, and Molecules

Matter is made up of tiny particles called atoms and molecules. These microscopic things combine to make up everything. Atoms are like building blocks that fit together to make different substances. Molecules are made up of two or more atoms that are held together by chemical bonds. Polymers are big molecules, made up of smaller molecules joined together.

Atoms are too small to see, but they are made up of even smaller particles inside.

Many of our everyday materials—like ice cream—are mixtures (a combination of different substances).

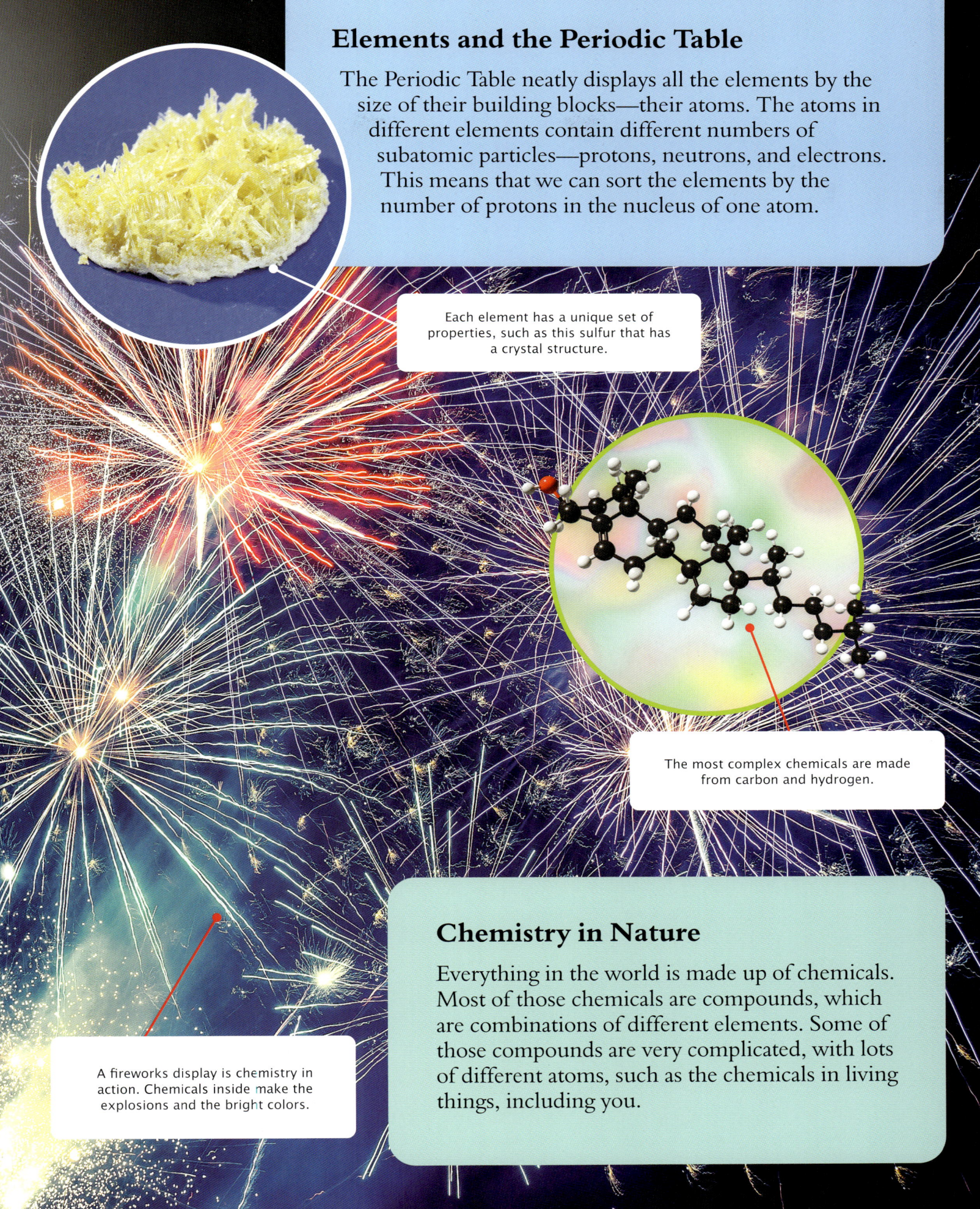

Elements and the Periodic Table

The Periodic Table neatly displays all the elements by the size of their building blocks—their atoms. The atoms in different elements contain different numbers of subatomic particles—protons, neutrons, and electrons. This means that we can sort the elements by the number of protons in the nucleus of one atom.

Each element has a unique set of properties, such as this sulfur that has a crystal structure.

The most complex chemicals are made from carbon and hydrogen.

Chemistry in Nature

Everything in the world is made up of chemicals. Most of those chemicals are compounds, which are combinations of different elements. Some of those compounds are very complicated, with lots of different atoms, such as the chemicals in living things, including you.

A fireworks display is chemistry in action. Chemicals inside make the explosions and the bright colors.

Chapter 1: Matter, Atoms, and Molecules

States of Matter

Matter is any substance that takes up space. It exists in three forms ("states" or "phases"), which are solids, liquids, and gases. All substances are in one of these three states at any time.

The sea is a liquid. It will take the shape of a bucket, but it will not expand to fill it.

Properties

Solids, liquids, and gases have different properties—they behave differently. You can hold a solid, and it won't change shape. Liquids run through your fingers, while gases can't be picked up at all. Scientists call both liquids and gases "fluids" because they flow and change shape to fit their containers. Shape and flow are properties of matter. Other properties are mass (the amount of matter), volume (the space it fills), density (heaviness), and compressibility (squashiness).

Oil floats on water because water is denser (heavier) than oil. Honey is denser than water. You can stack liquids of different densities in a jar, like this.

A pump compresses (squashes) air into a container. Gases have no definite volume, and they can be compressed. Solids and liquids are not easily compressed.

Properties of Mass, Volume, and Density

Solids and liquids have a definite volume and density, and these don't change, even when a liquid flows into a container. But when gases flow, they expand. The gas in a balloon becomes the shape of the balloon, and as more air is blown in, more pressure is put on the surface. If the balloon pops, the escaping gas flows out into the air, the mass of gas spreads out, and its volume and density change.

10

HALL OF FAME:
Maria the Alchemist
First Century CE

Maria the Alchemist was one of the first early scientists, or alchemists. She was already a historical figure by the fourth century when Zosimos of Panopolis quoted her work and described her as a "sage" (wise person). We think Maria invented some pieces of chemical apparatus, including the double-boiler for heating things gently. This was given the name "bain-marie" after Maria centuries later.

The air in an inflatable ball is a gas. When air is blown into the ball, it expands (spreads out) to fill the space.

Sand is a solid. It flows into a bucket, but importantly the size and shape of each tiny, individual grain of sand is solid and does not change.

Solids usually have a higher density than liquids, so they sink. But liquids are usually denser than gases, so solids that contain air will float.

A spade is a solid. Its shape and size don't change, unless it gets bent or broken.

DID YOU KNOW? Atmospheric pressure—the heaviness of air surrounding Earth—decreases with height. The change in pressure makes your ears pop in a plane.

Particle Theory

Solids, liquids, and gases behave differently. Solids have a shape, liquids flow, and gases escape in all directions. You can understand why, if you think of materials as made up of tiny, invisible balls, or particles. The particles' arrangement affects their properties (how they act). This is the particle theory of matter.

Solids and Liquids

In a solid, the particles are held tightly. They can't move around or change places. This gives the solid a definite shape and volume. It also makes it very dense (heavy) and hard to compress (squash). In liquids, the particles are held quite closely together, so liquids have a definite volume and are quite dense and hard to compress. But their particles are not neatly arranged. They can tumble past each other, so the liquid flows and changes shape.

Balls inside a ball pit are like the particles in a liquid. They are kept together, but they can move past each other. They flow to fit the shape of the ball pit.

Each ball is made of plastic. Plastic is a solid, so its particles are packed together, giving the ball a definite shape and size.

HALL OF FAME:
John Dalton
1766–1844

Dalton developed the idea of atomic theory. Other chemists wrongly believed that all particles (atoms) were alike. Dalton argued that the atoms of each element (the simplest chemicals) are alike, but that atoms of different elements are different. He also understood that when chemicals react, atoms rearrange into different compounds.

Gases and Diffusion

Gas particles are scattered and held together loosely. The wide spaces between them make the gas light and easy to compress. The particles have lots of energy to move around and spread out. This spreading is called diffusion. Gases have no definite volume. Their particles will diffuse until the gas has filled all the space it can, so a gas in a container will take on the shape of the container.

Cooking smells move through the house by diffusion. The delicious-smelling gas particles move freely through the air until they reach your nose!

Neatly stacked balls are like the particles in a solid. Each particle is held in place, and the substance has a definite shape and volume.

Balls escaping from the ball pit are more like particles in a gas. They are free to spread far apart.

DID YOU KNOW? Your body contains around 7 octillion atoms—that's 7,000,000,000,000,000,000,000,000,000!

Changing States

All materials are in one of the states of matter (solid, liquid, or gas), but they don't always stay in the same state. For example, when you freeze water into ice cubes and then leave the ice out to melt, the water changes from liquid to solid and back into liquid. Afterward, the water is the same as before.

Physical Change

Changing state affects how the tiny particles inside materials behave. The particles can be atoms or molecules (a group of two or more atoms). In a solid, they are held together by strong forces (bonds). These loosen when the solid melts—that's why melting ice loses its shape. The particles in a liquid are held by looser bonds. These loosen more when the liquid evaporates (changes into a gas).

The dry ice used in special effects is frozen carbon dioxide. It turns back into gas so quickly that it misses out the liquid state. When a solid turns directly into a gas, it is called sublimation.

Temperature

The state of a material is controlled by temperature (how hot or cold it is). Materials need heat to change to a state where their particles move around more. Gas particles move more than liquid particles, and liquid particles move more than solid particles. So, solids melt and liquids evaporate when they are heated. The heat gives the particles more energy and weakens the bonds between them. Going the other way, gases condense and liquids freeze when they are cooled.

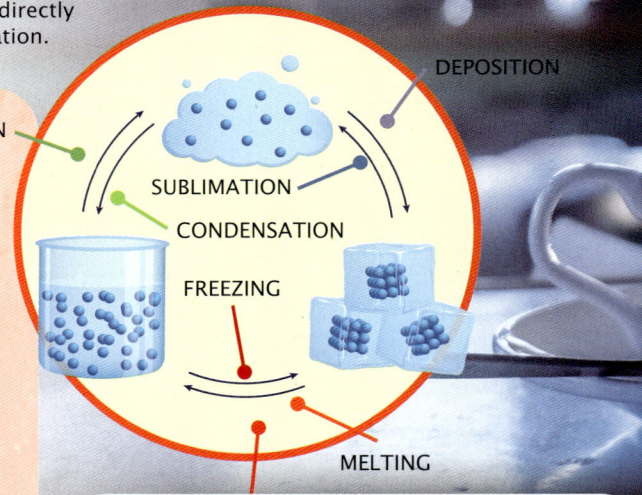

EVAPORATION · DEPOSITION · SUBLIMATION · CONDENSATION · FREEZING · MELTING

Freezing point and melting point are the same—the temperature where liquid turns into solid, or solid into liquid. Boiling point is the temperature where a liquid evaporates into gas, or the gas condenses into liquid.

DID YOU KNOW? Only one metal—mercury—is liquid at room temperature. It melts at a chilly −38.8°C (−37.9°F), compared with gold's 1,063°C (1,945°F).

The steam from a boiling kettle is water vapor escaping into the air.

Packed snow takes longer to absorb heat from the air, so a snowman melts more slowly than surrounding snow. Snow (water) has a melting point of 0°C (32°F).

Cooling water vapor condenses out of the air. It might run down the inside of the kettle or fog up the window.

The large bubbles of gas in the boiling kettle are water turning into water vapor (evaporating).

HALL OF FAME: Aristotle
384–322 BCE

Aristotle was an ancient Greek philosopher (deep thinker) who believed that all matter was made up of four "elements"—earth, air, fire, and water. His ideas led to the belief that metals could change into gold (alchemy) and they influenced experimental chemistry for centuries.

Mixtures and Solutions

Do you like mashed potatoes or mashed turnips? Or a mixture? You might enjoy it, or you might want to take one ingredient out! In chemistry, pure substances are like the potatoes (or the turnips) on their own. Impure substances (mixtures) are like the two together.

Pure and Impure Substances

Chemically pure substances contain only one thing. It could be one element (the simplest chemicals) like gold, or one compound (a chemical made of other chemicals joined up) like table salt. You can't take anything out of a pure substance without changing it chemically. Impure substances are a mixture of elements or compounds. Importantly, the elements or compounds in the mixture are not chemically joined up, so they can be separated.

Air is a mixture of gases, including oxygen, nitrogen, and carbon dioxide.

This cool, carbonated drink is a mixture of liquid juice, solid ice cubes, and carbon dioxide gas. The carbon dioxide escapes as the fizzy bubbles rise to the top.

Mixing

If you shake peas into a bowl of water, you can still see the peas. But you can't always see the things in a mixture. When you stir sugar into tea, the sugar dissolves—the bonds between the sugar molecules (particles) break. The sugar seems to disappear, but it doesn't—you can taste the sweetness. When something dissolves in something else it makes a mixture called a solution. We can mix different combinations of solids, liquids, and gases.

DID YOU KNOW? "Pure" fruit juice has no added sugar, but it's not chemically pure, because it contains many compounds.

Salt is soluble because it dissolves in water. Salt is the solute (the solid that dissolves).

The water is the solvent (the liquid that something is dissolving in).

The salt-water mixture is a solution. Adding more salt makes the solution more concentrated. Adding more water makes it more diluted.

The metal pan is insoluble (it doesn't dissolve in water).

HALL OF FAME:
Mary Elliott Hill
1907–1969

Hill was a chemist and teacher who was one of the first African American women to be awarded a master's degree in chemistry. She encouraged students to study and teach chemistry despite their social difficulties. She and her husband worked together on making soluble compounds used in plastics production.

Atomic Structure

Matter is made up of building blocks called atoms. Thinking of atoms as tiny balls helps explain how solids, liquids, and gases behave—but atoms also contain smaller, "subatomic" particles—protons, neutrons, and electrons. Different elements (the simplest chemicals) have different numbers of subatomic particles.

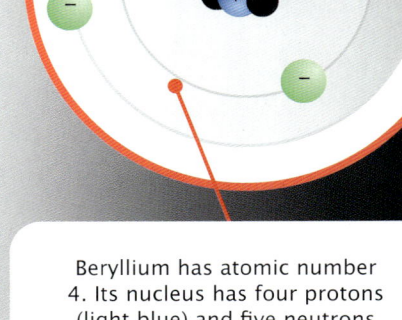

Beryllium has atomic number 4. Its nucleus has four protons (light blue) and five neutrons (dark blue). Four electrons (green) are arranged in two surrounding energy shells.

The Atomic Nucleus

The nucleus is in the middle of an atom. Protons and neutrons, which are almost 2,000 times heavier than electrons, are packed together inside the nucleus, so it is "massive" even though it's too small to see. Neutrons have no charge, and protons have a positive charge, making the nucleus positively charged. An element's atoms all have the same number of protons—this is its "atomic number."

Energy shells are stacked inside each other like Russian dolls. But the shells are not solid—only electromagnetic attraction holds the electrons close to the nucleus.

Electrons

Outside of the nucleus, the atom is mostly empty space. Electrons—very tiny particles with almost no mass and a negative charge—whiz around in energy "shells" (layers) surrounding the nucleus. Each shell can hold a set number of electrons, so atoms with more electrons have more shells. The equal and opposite charges of the protons and electrons attract each other—this electromagnetic force holds the electrons inside the atom. Atoms have the same number of electrons as protons, so the whole atom has no charge.

DID YOU KNOW? Protons and neutrons contain even tinier particles called quarks and gluons. Scientists know of 36 subatomic particles so far!

The total number of protons and neutrons in the nucleus is the atom's "mass number."

 Carbon 12
6 Protons
6 Neutrons
6 Electrons

 Carbon 13
6 Protons
7 Neutrons
6 Electrons

 Carbon 14
6 Protons
8 Neutrons
6 Electrons

Isotopes are forms of an element with different numbers of neutrons. A "normal" carbon atom nucleus—"Carbon 12"—has six neutrons, but other isotopes have seven or eight.

The first energy shell is closest to the nucleus. It can contain up to two electrons.

As the shells fill up with electrons, more shells are added. The heaviest atoms have over 100 electrons, in seven shells.

Bigger atoms have more shells, farther out from the nucleus. The second and third shells can contain up to eight electrons each.

HALL OF FAME: Joseph John Thomson
1856–1940

English physicist J. J. Thomson discovered the electron in 1897. His experiments showed that cathode rays—rays seen when electricity flows through gases at low pressure—were streams of particles with much less mass than atoms themselves. We now call these particles "electrons." He was awarded the Nobel Prize for Physics in 1906.

Elements

Each element is made up of just one kind of atom. The atoms of different elements have different numbers of subatomic particles—protons, neutrons, and electrons—inside their atoms. These differences give the elements their individual properties (how they look and behave).

We often see gold, silver, and copper made into decorative items, but these metallic elements have many more uses than trinkets.

Names and Numbers

We know of 118 elements. Around 90 are found naturally, while the rest are made by scientists and are usually unstable and quickly decay (break up). Every element has a name and a symbol of one or two letters. For example, hydrogen has the symbol H, and lead has the symbol Pb. Each element is also identified by its atomic number—the number of protons inside one atom.

Air is a mixture of gases, including the elements nitrogen, oxygen, and argon. The element in an airship—helium—is a gas that is lighter than air.

All Our Resources

At room temperature, only two elements are liquids—mercury and bromine—while 11 are gases. The rest are solids, and most of them are metals. Some elements—like nuggets of gold—occur naturally in pure form, but most are found in impure form, as compounds with other elements. Elements and their compounds make up the minerals and rocks of the Earth's crust.

Silicon oxide—a compound of oxygen and silicon—makes up most of Earth's crust as rocks or sand. Aluminum, iron, and calcium are the next three most common elements in the crust.

HALL OF FAME: Ida Noddack née Tacke 1896–1978

The element rhenium was first isolated in 1925 in Germany, by Ida Tacke, Otto Berg, and Walter Noddack, Ida's future husband. Rhenium has a very high melting point and is now used in aircraft engines. Ida was also the first to suggest that atoms bombarded by neutrons might split into smaller atoms. Four years later, nuclear fission was shown to be possible.

Copper is a "trace" element—a tiny amount in your body keeps you healthy.

Gold is very unreactive—it does not react with air and tarnish (go dull). It is used in medicine and in tooth fillings, and also in some electronic devices.

Silver is an excellent conductor (carrier) of heat and electricity. It reflects light well and is used in mirrors and solar panels.

DID YOU KNOW? The most recently discovered element, tennessine, was made in a laboratory in 2010. Very few atoms were made, and they decayed quickly.

Molecules

Atoms like to stick together! Only a few—the noble gases—hang around on their own. Most of them bond together with other atoms to make molecules. Molecules can be two atoms, or larger groups, and sometimes they are giant molecular structures. A crystal is a structure where the atoms or molecules join up in a regular, repeating pattern, often shown off in gems like diamonds.

A pencil drawing is made of graphite. Both graphite and diamond are giant molecular structures of carbon.

Diatomic Molecules

There are two atoms in diatomic molecules. If the atoms are identical, the molecule is "homonuclear." Elements with atoms that pair up in this way are "diatomic elements." The bond is made by sharing electrons, which fill up the atoms' energy shells with the most electrons they can hold. A hydrogen atom has one electron, but its shell can hold two—so two hydrogen atoms share their two electrons.

Hydrogen is a diatomic element. By pairing up and sharing their electrons to make a bond, two hydrogen atoms make a stable, homonuclear molecule.

HALL OF FAME: Rosalind Franklin 1920–1958

British scientist Franklin studied molecules, using X-ray crystallography. She helped to show he double helix (spiral) structure of the biological molecule, DNA (deoxyribonucleic acid). She also made important discoveries about the structure of viruses, as well as about the different forms of carbon in coal and graphite. Her work on carbon paved the way for the development of useful carbon fiber technologies.

DID YOU KNOW? The largest uncut diamond found, the Cullinan, measured 10.1 x 6.35 x 5.9 cm (4 x 2.5 x 2.3 inches). It was cut into over 100 gemstones.

Allotropes

The crystals of some elements are simple—they only contain one kind of atom—yet surprising. Their atoms join together in different ways to make different allotropes. Two allotropes of carbon are diamond and graphite (pencil "lead.") Diamond and graphite seem like different elements, because the way their atoms are arranged makes them look and behave very differently.

Diamond is the hardest natural substance on Earth. The best crystals can be cut into costly gemstones, but diamond is also used in industrial tools.

Diamond	Graphite	Fullerene
Tetrahedral	Trigonal planer	Spherical

The molecular sheets in graphite are weakly bonded, so graphite is softer than diamond, where bonds are strong in all directions. Another allotrope—buckminsterfullerene—has atoms in a sphere.

Graphite looks gray and dull. It is soft and used in pencils—its molecules slide over each other and easily rub off to leave a mark on paper.

Crystals of the element sulfur can be shaped like four-sided pyramids, or as long needles. The different forms are allotropes.

23

Compounds

Compounds are molecules with more than one kind of atom. They are made when the atoms of different elements react and bond together. The different chemicals in a compound can only be separated by a chemical change.

Compound Properties

No atoms are lost when chemicals react. This means the chemicals at the start (the reactants) and at the end (the products) contain the same atoms in different combinations. The products have new properties—they look and behave differently to the reactants. When you drink water, you are drinking a compound of hydrogen and oxygen, which are both gases in the air. And when you lick salt, you are eating a compound of chlorine (a gas) and sodium (a metal).

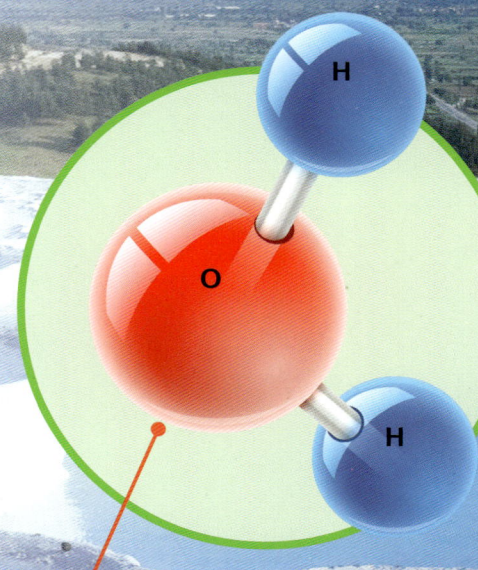

Chemicals have more than one name. A water molecule has two hydrogen atoms and one oxygen atom, so you could call it dihydrogen oxide!

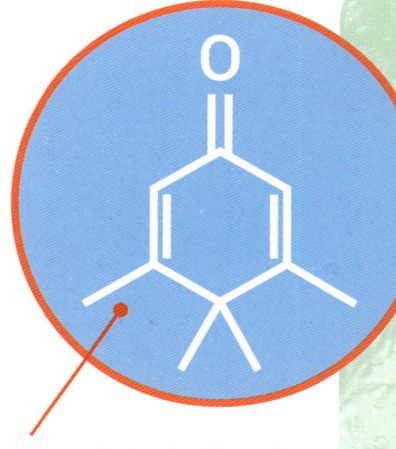

The structural chemical formula of 3,4,4,5-tetramethyl-2,5-cyclohexadien-1-one looks a bit like a penguin. Its common name is penguinone.

Names and Formulae

Some compounds contain many elements and have complicated names to describe them. Luckily scientists give chemicals simpler common names too. They also have a clever, short way of describing compounds—chemical formulae. Every element has a symbol of one or two letters, and these make up the chemical formulae of all possible compounds. The formula for water is H_2O, and this shows that the molecule has one oxygen and two hydrogen atoms. Another kind of formula, drawn to show the links between atoms, is the structural chemical

DID YOU KNOW? Made in 2014, the largest molecule, PG5, contains 17 million atoms of carbon, nitrogen, and oxygen. At 10 nanometers, it's as big as a virus.

The terraces of Pamukkale in Turkey are a natural wonder formed by underground thermal springs rising to Earth's surface.

The beautiful white limestone pools are made by calcium carbonate ($CaCO_3$), a compound of calcium, carbon, and oxygen, carried up in the water from underground.

Mixed with the hot water underground, $CaCO_3$ is in solution. When the water cools at the surface, the dissolved $CaCO_3$ turns back into a solid.

The process of a dissolved chemical leaving the solution and becoming solid is "precipitation."

HALL OF FAME:
Marie-Anne Lavoisier
1758–1836

Marie-Anne married Antoine Lavoisier when she was a young teenager. They had a laboratory at home in Paris, where they invited other scientists to watch and debate their experiments. As Antoine's coworker, illustrator, translator, and assistant, Marie-Anne was essential to their research. Among other things, they identified oxygen and showed that it reacts with other elements to make compounds.

Covalent Bonds

An atom is a nucleus orbited by electrons in energy shells. Each shell can hold a certain number of electrons and must be full before another shell can be added. When atoms react and join up to make a molecule, a bond is made by the electrons in their outermost shells. How many bonds an atom can make is its "valency."

The Shared Bond

Atoms are most stable (unreactive) when their shells are full, so you can think of atoms as "wanting" to fill up with electrons. They can do it by "sharing" their electrons, and this makes a covalent bond. The first shell of an atom holds up to two electrons. This explains why the two smallest atoms, with only one shell—hydrogen and helium—behave differently. A hydrogen atom, with one electron, readily shares it with another atom. But a helium atom has two electrons, so it doesn't react because it is already stable.

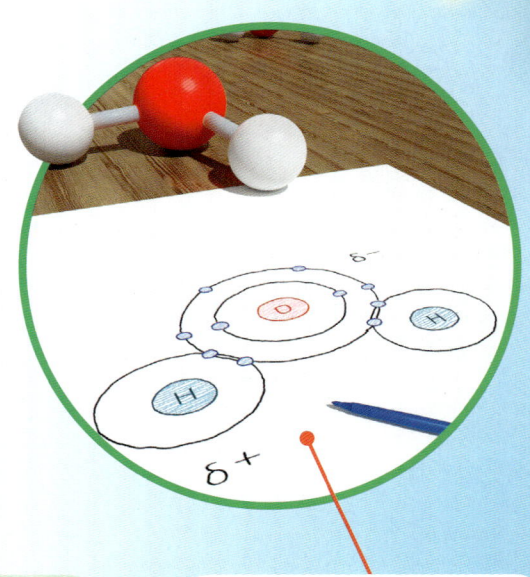

In a water molecule, the electrons donated by the oxygen atom fill up the hydrogens' single shells. The electrons donated by the hydrogen atoms fill up the oxygen's second shell.

Double Bonds

A pair of electrons shared between two atoms makes a single covalent bond. Some atoms can also make a double bond, by sharing two pairs of electrons. They can do this because the second and third shells can each hold up to eight electrons. An oxygen atom has six electrons in its second shell, so it "wants" two more (it has a valency of two). Two oxygen atoms can each donate (give) two electrons to share between them, making eight electrons for each of their outer shells.

Four electrons shared by two oxygen atoms create a double bond in an oxygen molecule. The nuclei, with eight protons and eight neutrons each, remain unchanged.

HALL OF FAME: Linus Carl Pauling
1901–1994

Pauling, born in Oregon, USA, was one of the first scientists to use quantum physics—where subatomic particles can behave like waves—to describe how atoms make bonds in molecules. He was awarded the Nobel Prize in Chemistry for this work in 1954. He went on to be awarded the Nobel Peace Prize in 1962 for his efforts to ban nuclear weapons tests and end the threat of nuclear war.

Candles are made of paraffin (petroleum) wax, which is the fuel when the candle is lit.

The hydrogen and carbon atoms in the candle wax make new covalent bonds with oxygen, producing carbon dioxide gas and water vapor.

The candle burns, producing heat and light, until all the wax is used up. The reaction is combustion—the burning of a fuel in oxygen.

Wax is a hydrocarbon—it has hydrogen and carbon atoms. When the candle burns, the heat melts the wax, which then turns into a vapor. Its molecular bonds are broken.

DID YOU KNOW? In quantum physics, it seems an electron can spin in two directions at once and can affect another electron even across the galaxy. Weird!

Ionic Bonds

Atoms stick together when an electrical force made by the electrons in their outermost energy shells makes a bond between them. In covalent bonds, the electrons are shared, but in ionic bonds the electrons move from atom to atom.

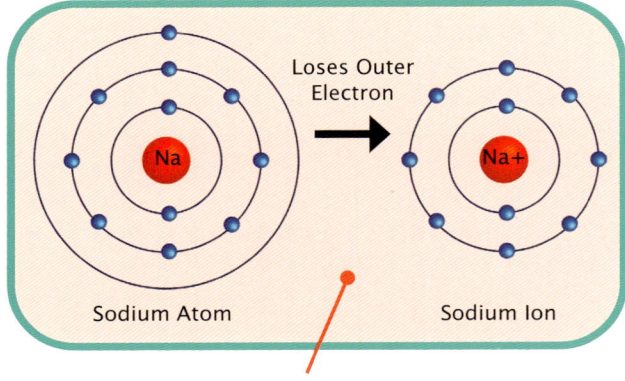

A sodium atom has only one electron in its outer (third) shell. It readily gives it up, because then its second shell becomes its outer shell, and it has the full, stable number of eight electrons.

Ions

An atom usually has no charge—the negative charge of its electrons balances out the positive charge of its nucleus. When an atom loses or gains an electron, it loses or gains some negative charge and becomes a charged particle called an "ion." If an electron moves away from an atom, it leaves an ion with a positive charge—a "cation." An atom that gets an extra electron becomes a negative ion—an "anion."

Ionic Lattices

The opposite charges of cations and anions attract each other by electrostatic force, and this makes an ionic bond. Compounds containing ionic bonds are crystals with a regular, repeating pattern of negative and positive ions in an ionic lattice structure. The electrostatic forces act in all directions, making ionic lattices strong. The compounds have no overall charge, because the negative and positive charges balance each other out. They usually have high melting and boiling points, because a lot of energy is needed to break all the ionic bonds.

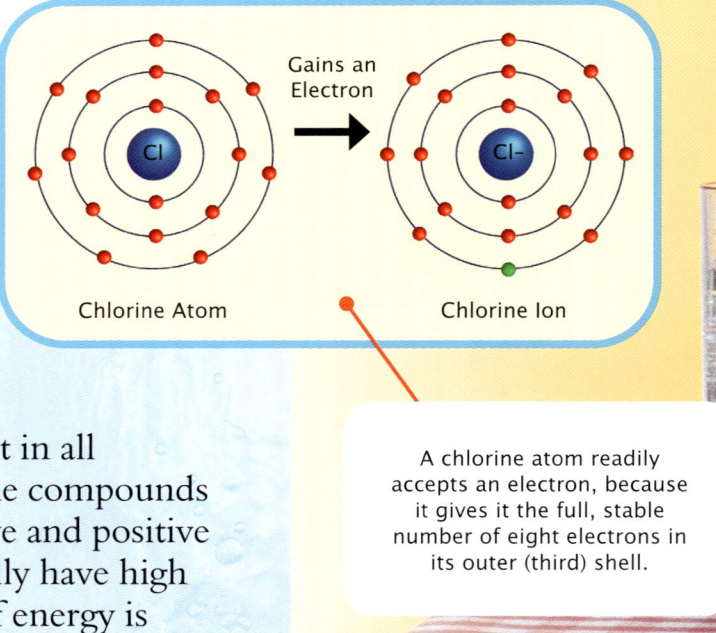

A chlorine atom readily accepts an electron, because it gives it the full, stable number of eight electrons in its outer (third) shell.

DID YOU KNOW? Our nerve cells use sodium and other ions to send electrochemical messages that travel at up to 288 km/h (180 mph)!

Reactions

When substances react together, the chemicals before the reaction (the reactants) make new chemicals (products) with new properties. Their atoms rearrange—they break their bonds and make new ones. No atoms are lost, so the same atoms are combined differently in the reactants and the products.

> Atoms bond together like a group of dancers. They can hold hands (bond), and they can let go, move around, and find new partners.

Chemical Equations

A word equation describes a reaction. For example, the word equation for the reaction of carbon and oxygen to make carbon dioxide is:

carbon + oxygen ⟶ carbon dioxide

The chemical equation tells us more. It uses chemical symbols and formulae to show the type and number of atoms in the molecules of the reactants and products:

$C + O_2 \longrightarrow CO_2$

It shows that one atom of carbon reacts with two atoms of oxygen to make a carbon dioxide molecule containing one carbon and two oxygen atoms. The equation is "balanced," with three atoms before the arrow and three atoms after it.

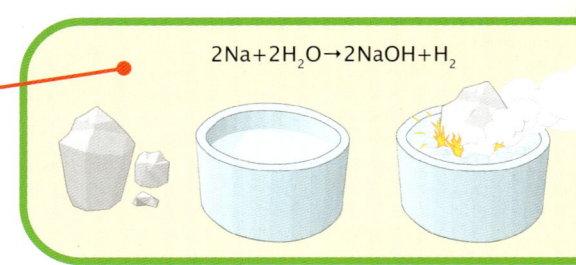

This equation shows that two atoms of sodium and two molecules of water produce two molecules of sodium hydroxide and a molecule (two atoms) of hydrogen.

$2Na + 2H_2O \rightarrow 2NaOH + H_2$

HALL OF FAME: Antoine Lavoisier 1743–1794

Lavoisier developed a theory of chemical reactivity based on experiments. He showed that as much matter exists after a reaction as before, and in 1789, he published a book stating the Law of the Conservation of Mass for the first time. The book also introduced a new way of naming compounds that is still the basis of chemical names today. Lavoisier was executed during the French Revolution.

When copper wire is suspended in a clear silver nitrate solution, copper atoms take the place of silver and change the solution to blue copper nitrate. Displaced silver forms crystals on the wire.

Reactivity

Reactivity is how readily a chemical reacts with others. Some metals, like sodium, are so reactive that they are only found as compounds in nature. Others, like gold, are not at all reactive. In the reactivity series of metals (a list of metals from most reactive to least reactive), sodium is more reactive than copper, which is more reactive than silver, which is more reactive than gold. A more reactive metal will replace a less reactive metal in a compound in solution. This is displacement.

The members of the group stay the same, even when they change partners.

Some members change partners more easily than others.

DID YOU KNOW? Chemical reactions keep us alive—that's biochemistry. Trillions of biochemical reactions are happening in your body right now.

Chemical Reactions

In a chemical reaction, the reactants' atoms rearrange to make the products. Some reactions are reversible; for example, nitrogen dioxide gas breaks down into nitrogen monoxide and oxygen when heated, and changes back when cooled. Other reactions, like burning and corrosion, are irreversible—it's impossible to get the original reactants back.

Cooking and Combustion

A cook can't unbake a cake, because the ingredients have been permanently, chemically changed. Baking a cake involves reactions such as thermal decomposition, which is breaking apart (rather than burning or melting) a substance by heat. Combustion—when a fuel is burned in air—is another useful, irreversible reaction. The fuel, such as wood, reacts with oxygen in the air to release energy in the form of light and heat—but we can't turn the ash back into wood afterward.

Color can neatly demonstrate a reversible change—a flask of nitrogen dioxide loses its color when heated, and returns to red-brown when cooled.

HALL OF FAME:
William Jacob Knox
1904–1995

Knox earned his Chemistry PhD at Massachusetts Institute of Technology (MIT). During World War II, he was a supervisor in the Manhattan Project, using the corrosive gas, uranium hexafluoride, to separate isotopes. Later, he became only the second Black chemist to work at the Eastman Kodak company. The lifelong racial prejudice he faced inspired him to fight for civil rights, and he set up scholarships to help minority students.

DID YOU KNOW? The planet Mars is rusty—it looks red because iron in its soil reacted with oxygen billions of years ago, when the planet had liquid water.

Corrosion and rusting

Corrosion is an oxidation reaction in which a metal oxidizes—gains oxygen from the air. It is an example of permanent change and can be a problem in buildings, because the metal gets weaker as it continues to change into the metal oxide. Rusting is the type of corrosion that happens when iron is exposed to oxygen in air or water.

Iron reacts with oxygen to produce iron oxide—rust.

The statue is made of iron, covered with copper sheets. Copper reacts with oxygen in the air to form copper oxide.

The copper oxide has reacted with carbon dioxide, sulfur, and salt in the air to create the copper compounds that make the blue-green pigment, verdigris.

The Statue of Liberty, with its pedestal, is 93 m (305 ft) high. It has stood on Liberty Island, USA, since 1886. The blue-green statue was once bright red-brown, like a new copper penny.

The verdigris keeps the copper underneath from reacting more—so the statue is protected by the corrosion on the surface.

Exothermic and Endothermic Reactions

When chemicals react, their atoms are rearranged—their bonds are broken and new ones made. Energy is taken in when bonds are broken, and released when they are formed. The difference in the energy absorbed and released by all the bonds makes the total reaction endothermic or exothermic. The energy is usually (but not always) heat.

Nitric acid is a compound of hydrogen, nitrogen, and oxygen. It is a strong oxidizing chemical (it provides oxygen for reactions), and it can produce explosions—violent exothermic reactions.

Exothermic Reactions

Exothermic reactions release energy overall, so they can feel warm. Many oxidation reactions—where a substance gains oxygen—are exothermic. Disposable hand warmers use a surprising exothermic oxidation reaction—rusting! The pouch contains separated water and iron powder. When it's activated, the iron reacts with oxygen in the water, producing iron oxide (rust) and heat. Respiration is another exothermic reaction. It produces energy for your body to use.

When you snap a glow stick, you start an exothermic reaction that releases energy in the form of light.

DID YOU KNOW? Composting is exothermic. Australian brush turkeys don't sit on their eggs; they build a compost pile of decaying vegetation that heats up.

As well as energy, the products of combustion are carbon dioxide and water. Bonfire smoke also contains carbon monoxide and other chemicals from the fuel.

Combustion is an excellent example of an exothermic reaction. It releases lots of energy in the form of heat and light.

Endothermic Reactions

In endothermic reactions, more energy is absorbed than released, so they can cool the surroundings. Cooking and thermal decomposition are endothermic—your cake mixture needs to absorb heat, so that the ingredients can turn into a cake. Photosynthesis, where plants turn carbon dioxide and water into glucose and oxygen, is endothermic—plants absorb the sun's energy for the reaction.

The thermal decomposition of sodium bicarbonate (baking soda) in a cake mix means that the baking powder breaks down to produce carbon dioxide, which makes the dough rise.

We ignite (light) a bonfire with a source of heat, like a match, and then enjoy the fire as the fuel burns.

The fire will burn until all the fuel has been used up and only ash is left.

HALL OF FAME: May Sybil Leslie
1887–1937

Leslie perfected the manufacturing conditions for nitric acid, used in explosives production, in a government laboratory during World War I. She was an English chemist who worked for a time with the Nobel Prize winner Marie Curie, and studied the radioactive compounds of thorium and actinium. She was only 49 when she died, possibly due to radiation exposure.

35

Important Properties

Glass or transparent (see-through) plastic lets the light into a greenhouse.

Glass slippers and gingerbread houses are only good in fairy tales. In real life, we don't want brittle shoes or soggy houses! We choose materials with properties (qualities) that make them fit for the job. Hardness (how hard or soft something is), roughness, flexibility, and permeability (having pores that let liquid through) are some physical properties of materials.

Metals and Ceramics

Metals are good for building because they're hard and strong. They are also malleable (easily shaped), ductile (can be pulled into thin wires), and good conductors (carriers) of electricity and heat. Ceramics, such as porcelain and clay, are waterproof, strong, and not good conductors, but they are brittle—they break instead of bending. Ceramics are used for mugs, toilets, bricks, and car brakes.

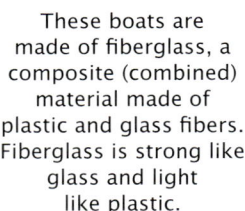

These boats are made of fiberglass, a composite (combined) material made of plastic and glass fibers. Fiberglass is strong like glass and light like plastic.

A rubber band stretches because its molecules are long and tangled. When we pull it, the molecules straighten up and get longer. The stretching is reversible—the molecules bounce back.

Plastics and Rubber

Plastics are polymers (giant molecules) made from fossil fuels. Rubber is an elastic (stretchy) polymer. Plastics may be hard or soft, flexible or stiff, transparent or opaque. Plastic comes in many forms. It is so versatile (useful in lots of jobs) that we rely on it all the time.

HALL OF FAME: Angie Turner King 1905–2004

King was the grandchild of slaves, and her father encouraged her education. She earned a master's degree in mathematics and chemistry at Cornell University and had a long career in science education. She taught chemistry to students at school and university, and even to soldiers during World War II. Many of her students had outstanding careers in science.

An apron may be made of waterproof cloth to keep the employee's clothes dry. A material that doesn't allow water to soak through it is "impermeable."

The plastic hose can bend around corners because it's flexible. The spray pieces are rigid (stiff) plastic or metal.

Plant pots are often made of a stiff, opaque (not see-through), recycled plastic. The large troughs may be made of wood, which is strong and permeable.

Ceramic tiles on the floor are smooth to walk on, hard-wearing, and easy to clean.

DID YOU KNOW? A diamond can scratch anything because diamond is the hardest natural substance on Earth. Diamond-edged tools are even used to drill rocks.

Chapter 2: Elements and the Periodic Table

The Periodic Table

The Periodic Table neatly displays all the elements by the size of their building blocks—their atoms. The atoms in different elements contain different numbers of subatomic particles—protons, neutrons, and electrons. This means that we can sort the elements by the number of protons in the nucleus of one atom.

Order of Size

The elements are placed in Periods (rows) and Groups (columns). Hydrogen, in the first square in Period 1, has only one proton and one electron. Helium is a bit bigger, with two protons and two electrons. Then comes lithium, which starts Period 2. The atoms get heavier along the Periods and down the Groups, until the final square in Period 7, Group 8. This is organesson, the most giant of atoms, with 118 protons.

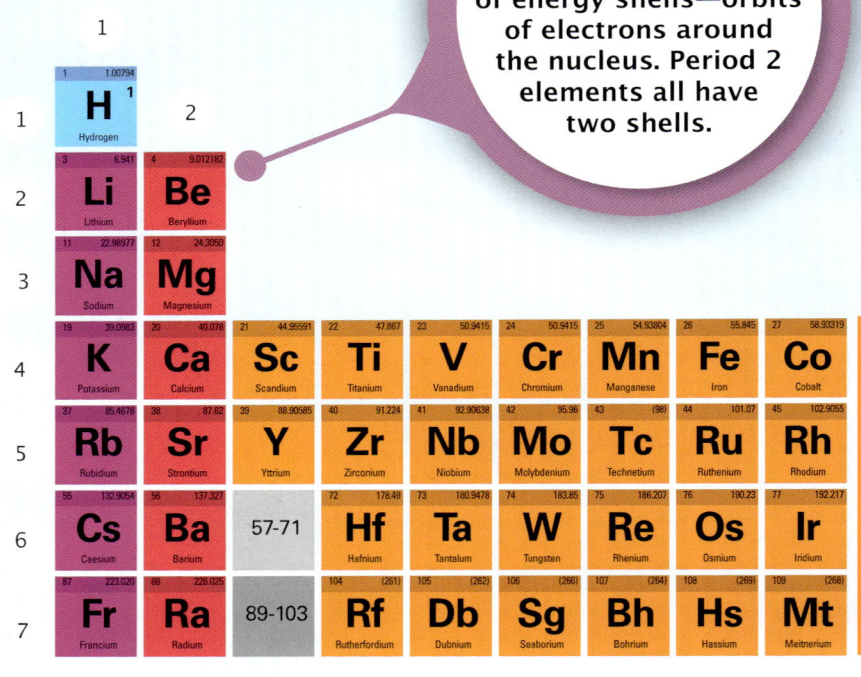

All the elements in one Period (row) have the same number of energy shells—orbits of electrons around the nucleus. Period 2 elements all have two shells.

The lanthanides and actinides are very similar elements squeezed in between squares 57 and 71, and squares 89 and 103.

Elements such as the transition metals and non-metals are also grouped together by their similar properties.

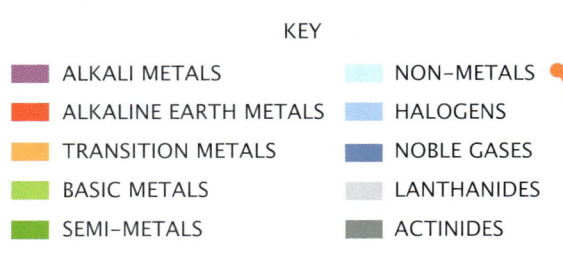

KEY
- ALKALI METALS
- ALKALINE EARTH METALS
- TRANSITION METALS
- BASIC METALS
- SEMI-METALS
- NON-METALS
- HALOGENS
- NOBLE GASES
- LANTHANIDES
- ACTINIDES

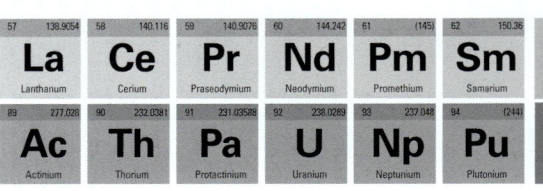

HALL OF FAME: Julia Lermontova
1846–1919

Julia Lermontova had to study in Germany because Russian universities wouldn't take women, and she was only the second woman worldwide to receive a chemistry doctorate. Her work on platinum and related metals was respected, and she helped Dmitri Mendeleev fill in the gaps when he put the elements in order to make the first version of the Periodic Table.

All the elements in a Group have the same number of electrons in their outer energy shells. This makes them look and behave in similar ways. Group 7 elements all have seven electrons in the outer shell.

What It Tells Us

One square gives the name and symbol of an element, with its "atomic number"—that is, number of protons (which is the same as its number of electrons). It also shows the relative weight of the atom—its "atomic mass"—which relates to the mass of its protons plus neutrons. Atomic mass has a decimal point because it's an average of different isotopes with different neutron counts. Other versions of the Periodic Table may give an atom's "mass number"—number of protons plus neutrons—which is always a whole number.

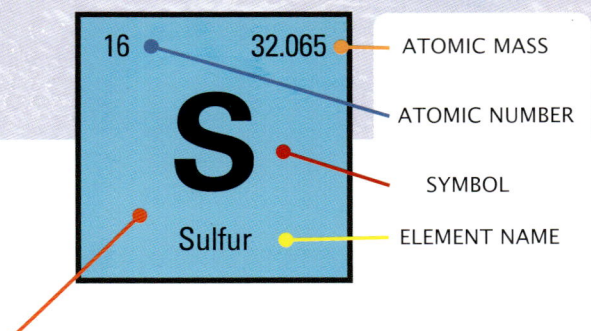

ATOMIC MASS
ATOMIC NUMBER
SYMBOL
ELEMENT NAME

Sulfur (S) has 16 protons in one atom. Its usual mass number is 32, showing that it has 16 neutrons (32 minus 16).

DID YOU KNOW? In the nineteenth century, people swallowed antimony (element 51) pills to treat constipation. The pill could be retrieved afterward and used again!

Element Groups

The Periodic Table ranks the elements' atoms in size order by number of protons, which is the same as their number of electrons. Atoms have energy shells that hold up to a certain number of electrons. The first (innermost) shell can hold two electrons. When it's full, a second shell outside of it holds up to eight electrons. Successive shells are farther out from the nucleus. Shells are most stable (unreactive) when they hold the most electrons they can.

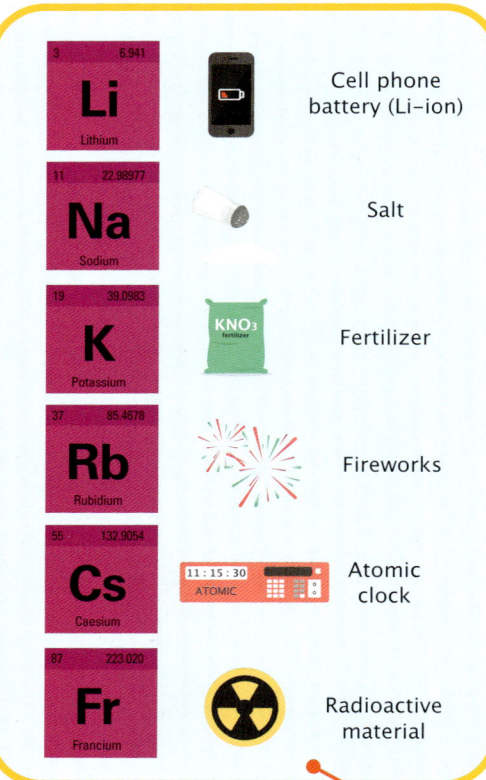

The Outer Shell Electrons

Elements in Periodic Table Groups (columns) behave alike, because of their pattern of electrons. For example, each Group 1 element (going down the Group) has one more energy shell than the one before—so lithium has two, sodium has three, and francium has seven shells—but they all have an outermost shell with just one solitary electron. This makes them behave alike, since they all react readily to give up the solitary electron. Elements in other Groups are also alike, because they have the same number of electrons in their outer shells.

All alkali metals react with water, giving off hydrogen gas and heat. The reaction gets more violent the farther down Group 1 the element sits.

Group 1 elements—the alkali metals—have similar properties, but they are used in lots of different industries.

HALL OF FAME: Marguerite Catherine Perey 1909–1975

In 1939, francium became the last naturally occurring element to be discovered, and it was the only element to be discovered solely by a woman. Marguerite Catherine Perey was separating radioactive elements when she found one that fit the gap at atomic number 87 in Mendeleev's Periodic Table. She named it francium after her home country, France.

Alkali Metals

The alkali metals are the Group 1 elements—lithium, sodium, potassium, rubidium, and caesium, plus radioactive francium. They are shiny metals, soft enough to cut with a knife. They all react readily with other chemicals, but the ones lower down the Group are more reactive than those above. They all react with cold water by releasing heat—an exothermic reaction. They react with some non-metals to form white, soluble, crystalline salts. For example, sodium reacts with chlorine to form sodium chloride—the salt in our food.

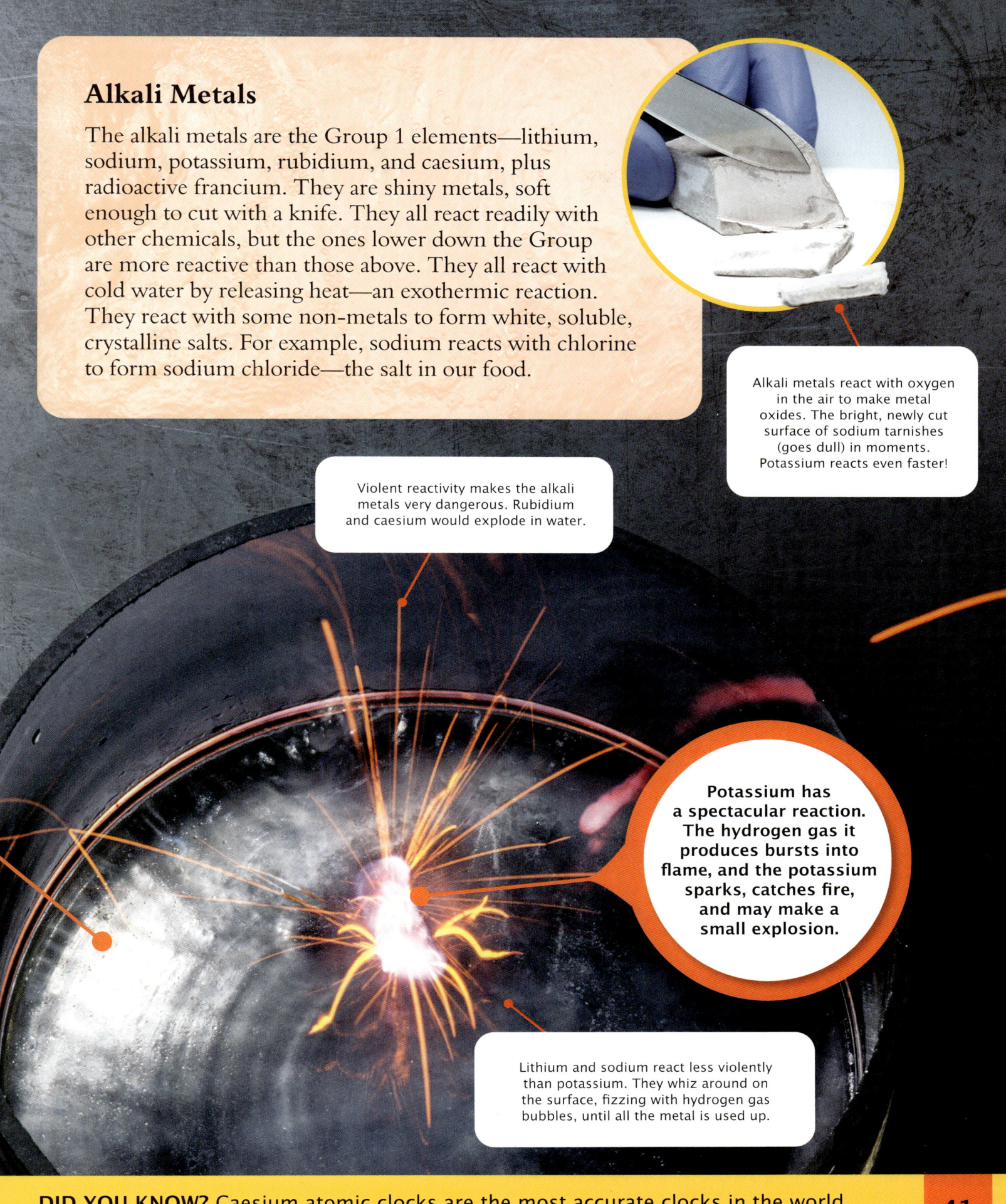

Alkali metals react with oxygen in the air to make metal oxides. The bright, newly cut surface of sodium tarnishes (goes dull) in moments. Potassium reacts even faster!

Violent reactivity makes the alkali metals very dangerous. Rubidium and caesium would explode in water.

Potassium has a spectacular reaction. The hydrogen gas it produces bursts into flame, and the potassium sparks, catches fire, and may make a small explosion.

Lithium and sodium react less violently than potassium. They whiz around on the surface, fizzing with hydrogen gas bubbles, until all the metal is used up.

DID YOU KNOW? Caesium atomic clocks are the most accurate clocks in the world, losing or gaining just one second in 1,400,000 years!

Tiny Hydrogen

The tiny atoms of hydrogen and helium were the first to form after the Big Bang. Hydrogen is the smallest atom, with one proton, one electron, and no neutrons. Hydrogen is often shown at the top of Group 1—but it's nothing like the solid, soft, shiny alkali metals of Group 1. Hydrogen is unique.

Highly Reactive

Pure hydrogen is rare on Earth. It is very reactive, so it's usually found in compounds with other elements. It is a nontoxic (not poisonous) gas, with no color, smell, or taste. It's the lightest element, used in weather balloons that gather information high in the atmosphere. It readily explodes with oxygen, and burns in air to produce water and energy. Hydrogen is so reactive because it easily shares or gives up its solitary electron.

Hydrogen gas must be pressurized (compressed) or liquefied to store and move. It becomes liquid when it's supercooled to −253°C (−423°F). This is expensive and can be dangerous.

Every shuttle flight to the International Space Station uses about 2,250,000 liters (500,000 gallons) of hydrogen liquid as fuel.

Powerful Hydrogen

We wouldn't be here without hydrogen. Life-giving water is made from hydrogen and oxygen atoms. With carbon and other atoms, hydrogen forms the organic chemicals that make up every living thing. Hydrogen gives us fuel via the hydrocarbon molecules of crude oil and natural gas. Pure hydrogen is also becoming more important as a renewable, clean fuel in vehicles and aircraft. Hydrogen takes us into space—it has been used as rocket fuel since the beginning of space exploration in the 1950s.

DID YOU KNOW? Hydrogen is the most abundant element in the Universe—its atoms make up over 70 percent of the total mass of matter.

HALL OF FAME:
Theophrastus von Hohenheim (Paracelsus)
1493–1541

The earliest chemists were called alchemists. Paracelsus was an alchemist who established the use of chemicals such as mercury and sulfur in medicines. He observed that poisons could also heal, at a different dosage. Paracelsus is believed to have discovered hydrogen without realizing it, when he noticed that iron filings in sulfuric acid produced gas bubbles that could burn.

As hydrogen-fueled cars are being improved, some roadside fuel stations have hydrogen pumps next to the more common gas pumps that use fossil fuels.

We need clean energy alternatives like hydrogen cars, because fossil fuels contribute to climate change and they will run out.

Compressed hydrogen gas is squashed into the vehicle's fuel tank. When it's fed into the fuel cells, it reacts with oxygen, producing energy which is turned into electricity.

Electricity runs the car's engine. The only other product is water, so there are no polluting emissions.

43

Alkaline Earth Metals

The alkaline earth metals—beryllium, magnesium, calcium, strontium, barium, and radium—are the elements in Group 2 of the Periodic Table. Their atoms have two electrons in their outer energy shell. This makes them very reactive, although not as reactive as the alkali metals in Group 1.

Increasing Reactivity

Like the alkali metals, the alkaline earth metals are silvery or gray, and their reactivity increases as you go down the Group. Beryllium, at the top, is the least reactive. It needs a very high temperature before it can react with water. Magnesium, the next element down the Group, fizzes a little in cold water, while the reaction is more and more vigorous for calcium, then strontium, and then barium.

Beryllium and magnesium are used in metal alloys (mixtures) in aircraft and cars because they are light. Barium, which is much heavier, helps doctors see inside patients' bodies with X-rays.

Magnesium helps the enzymes in our bodies to work, so it's important to eat plenty of magnesium-rich foods.

Alkaline Medicines

The alkaline earth metals get their name because their compounds make solutions with water that are "alkaline" (above pH 7 on the scale of acidity/alkalinity). Milk of magnesia is a suspension of magnesium hydroxide (H_2MgO_2). It's used as an alkaline medicine to treat indigestion because it neutralizes (cancels out) the stomach acids that cause the pain.

DID YOU KNOW? The hands of bedside clocks used to be painted with glow-in-the-dark paints containing radioactive radium, all the way up until the 1960s.

Alkaline earth metals are so reactive that they can only exist naturally as compounds. The gemstone emerald is a compound of beryllium. Small amounts of chromium make it green.

Marine snails make seashells from calcium carbonate. Other invertebrates like corals and crabs also build protective skeletons from calcium compounds.

Calcium is essential for all living things. Vertebrates, like us, use calcium compounds to build strong bones and teeth. The main compound in bones is calcium phosphate.

Pollutants such as sulfur dioxide make "acid rain." This softens the skeletons and shells of sea creatures, and is harmful to life both in the oceans and on land.

HALL OF FAME: Isabella Cortese

Sixteenth Century

Isabella Cortese was a well-traveled Italian alchemist who wrote the first book of cosmetic recipes, published in 1561. The book gave advice on running a household and how to make medicines and cosmetics, as well as discussing how metals might be turned into gold. It was very popular with the public and was republished several times.

The Halogens

That familiar swimming-pool smell is from compounds of chlorine, the main chemical used to kill germs and keep the pool clean. Chlorine is one of the halogens—non-metal elements in Group 7 of the Periodic Table. They are fluorine, chlorine, bromine, and iodine, all commonly used in disinfectants.

The chlorine in pool sanitizers breaks down in the water into hypochlorous acid (HOCl), a weak acid, and hypochlorite ion (ClO−).

Reactive Group 7

Halogen atoms only need one more electron to reach the maximum, stable number in their outer energy shells, so they are very reactive. They readily take an electron from another atom, and so become particles (ions) with a negative charge. Elements at the top of the Group are more reactive; fluorine's smaller size means the nucleus pulls more strongly on electrons of other atoms, making it the most reactive of the Group. It is one of the most reactive elements of all—it can make steel wool burst into flames.

When heated, iodine changes from the solid to a purple gas without becoming a liquid in between. This is sublimation.

HALL OF FAME:
Henry Aaron Hill
1915–1979

Henry Aaron Hill completed his PhD at Massachusetts Institute of Technology (MIT) and later became the first Black president of the American Chemical Society. He studied compounds used to make fluorine-containing plastics. Hill established companies supplying chemicals used in plastics production, and he offered research and consultation in polymer chemistry.

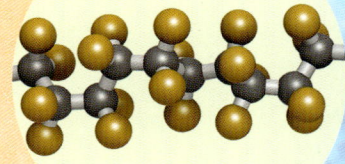

DID YOU KNOW? Your body contains about 3 mg of fluoride (fluorine compounds). Fluoride guards against tooth decay, and it's added to toothpaste.

Ribbon seaweeds (kelp) take in iodine compounds from seawater, so they are a good food source. Too little iodine in the diet can cause goiter (swelling of the thyroid gland).

Hypochlorous acid and hypochlorite kill the bacteria and other microorganisms that can cause stomach and ear infections.

Chlorine compounds can irritate skin, and hypochlorite makes fabrics fade—so always rinse your body and your swimsuit when you leave the pool!

Non-stick pans are coated with PTFE (polytetrafluoroethylene), a plastic made of carbon and fluorine. When it was invented in 1938, it was the slipperiest substance known.

Poisons with Useful Compounds

The halogens are strong-smelling, poisonous elements. Their atoms can bond in pairs to form molecules with two identical atoms, but they do not exist in these pure forms in nature. They combine with other elements as compounds in rocks and the oceans. They react with metals to form ionic salts called metal halides, such as sodium chloride (table salt). The halogens behave alike, but they do not all look alike. Fluorine and chlorine are greenish gases, bromine is a dark red, oily liquid, and iodine is a black solid.

Noble Gases

The last column in the Periodic Table is Group 8. The elements in this Group are the noble gases, so-called because they won't join a gang with other atoms to make compounds. They are helium, neon, argon, krypton, xenon, and radon. They have no color or smell, and—apart from radon, which is radioactive—they are safe to use.

Red neon lights contain pure neon. More "neon" colors are produced by the other noble gases. Neon lights are tubes containing the gases at low pressure.

Inert Gases

The noble gases are known as the inert gases because they're so inactive. They don't react because they have a full outer shell of electrons. This means that they have a valency (bonding ability) of 0—they can't make bonds because they don't need to share, borrow, or lend electrons with other atoms. They exist as single atoms, but they are naturally rare—except for argon, which makes up 1 percent of the air. We take it in with every breath, but it has no effect on our bodies.

Argon is used as a shielding gas in welding (joining metals). The equipment releases the gas around the metal as it melts, protecting it from oxygen and moisture in the air.

HALL OF FAME: Marie Curie
1867–1934

Polish scientist Marie Curie, and her husband Pierre, discovered the elements radium and polonium from the mineral pitchblende. In 1900, they observed that radium released a gas during radioactive decay. Another scientist, Friedrich Dorn, also observed the new radioactive gas, which was later named radon. The Curies were awarded a Nobel Prize in 1903, and Marie received a second one in 1911.

Helium and Neon

Our most familiar noble gases are helium and neon. Helium, lighter than air, is used in balloons, but it's also an important cooling agent in spacecraft and advanced research equipment such as the Large Hadron Collider. Neon is used as a powerful coolant and in electrical equipment. It gives off a red glow when an electric current runs through it, and it's commonly used in advertising. The barcode scanners in stores are helium-neon gas lasers.

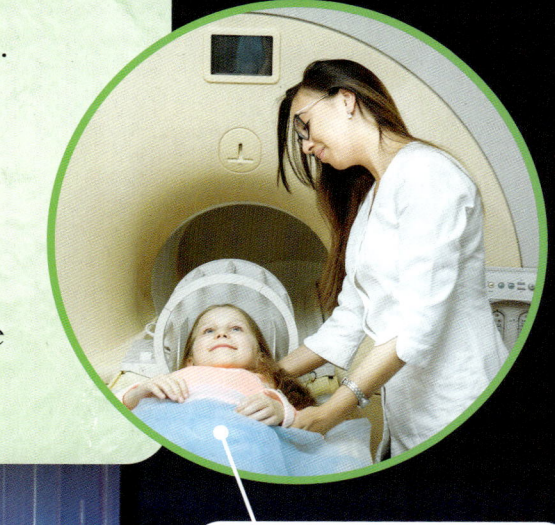

The excited electrons release energy as light. The different noble gases emit light of different wavelengths, which is why we see them as different colors.

Helium is used to cool the magnets of MRI scanners—hospital machines used for looking at details inside patients' body parts.

The gases produce the bright colors when an electric current runs through them and gives energy to the electrons in their atoms.

Old-fashioned incandescent light bulbs emit light from a heated tungsten filament. They are full of argon, which keeps the heated filament from reacting with oxygen in the air and corroding.

DID YOU KNOW? Nuclear reactors give off krypton. In the Cold War, levels of krypton-85 were used to track the secret building of nuclear weapons.

Metals

Metal elements are mainly strong, high-density, malleable (they can be shaped), and good conductors of heat and electricity, with high melting and boiling points. They usually react with oxygen to form oxides that are basic (alkaline, not acidic), and they react with acids to make a metal compound called a "salt", plus hydrogen. They lose electrons in reactions to form positive ions (cations).

Metals are good conductors of electricity because of their metallic bonds. The outer electrons are loosely bonded and can flow between the atoms, carrying the charge through the metal.

Transition and Post-transition Metals

The transition metals, which fill the central panel of the Periodic Table, are "typical" metals—hard, heavy, and shiny. They are less reactive than the highly active alkali metals and alkaline earth metals. Iron, a transition metal, is attracted to magnets. Iron alloys (mixtures) are ferrous metals and are also magnetic. The post-transition metals, or "basic" metals, include aluminum and lead. They are softer than the transition metals and have lower melting points.

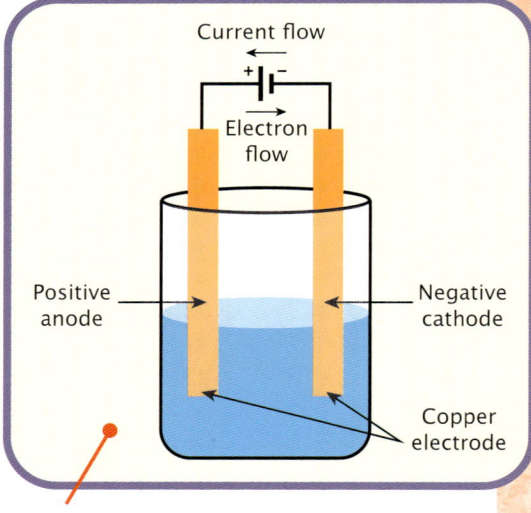

Electrolysis separates metal compounds that are dissolved or molten. Here, the electric current separates copper (Cu) from copper sulfate solution.

Separating Metals

Metals occur naturally in rocks as ores—metal oxides and other compounds. They can be separated by electrolysis and other methods. Some metals, including zinc, iron, and copper, can be extracted using carbon. Carbon is a non-metal that is more reactive than those metals, so it can "displace" them—remove them from their oxides and take their place, leaving pure metal behind. That's why carbon is often included in the metals' reactivity series—the list of metals from most reactive (potassium) to least reactive (gold).

HALL OF FAME: Jabir Ibn Hayyan 721–815

Jabir Ibn Hayyan was a Muslim alchemist born in Iran and is known as the father of Arabic chemistry. He developed orderly ways to experiment and analyze substances, and he influenced theories of chemistry and modern pharmacy. He is thought to have written hundreds of works describing chemical methods that included making alloys, and purifying and testing metals.

Industrial sorting claws may use magnets to separate ferrous metals such as steel from nonmagnetic materials such as aluminum and plastics.

All used metals should be recycled to avoid waste and reduce the need for mining new ores. Metals are easily recycled, but first the waste must be sorted and separated.

Alloys are often more useful than pure metals. Steel is an alloy of iron with carbon and other elements. It's stronger and lighter than iron, and is used in cars and buildings.

Stainless steel is made by adding chromium, which protects it from corrosion and rusting.

DID YOU KNOW? The metal bismuth repels magnets. So a magnet placed between an upper and lower block of bismuth will float in the air between them!

Non-metals and Semi-metals

There aren't many non-metal elements, but they are vital to life—from the carbon that builds living cells to the oxygen we breathe. Most of the Periodic Table is taken up by the metals, with the non-metals on the right of the Table. Around the zigzag line between them are the semi-metals, or "metalloids."

Semiconductors

Semi-metals—materials that sometimes conduct electricity—can be used to make semiconductors. The semi-metal silicon is like a metal because it's shiny and has a high melting point, and it's also like a non-metal because it has a low density and is brittle. It can conduct electricity in the right conditions. A pure silicon crystal can't conduct electricity because its electrons are tightly bonded. But when atoms of an impurity, such as arsenic, are added to it, an electric current can flow. This is "doping," and it allows all our electronics devices to be built around silicon.

Doping silicon with arsenic gives extra, free electrons, which carry a negative charge. Doping with indium makes spaces without electrons. The spaces move as electrons flow into them, so they carry a positive charge. Both types used together make a switch.

HALL OF FAME:
Esther M. Conwell
1922–2014

Esther M. Conwell was an American chemist and physicist whose love of puzzles helped her explain semiconductors. She described how electrons flow through semiconductors in the Conwell–Weisskopf theory, and the breakthrough revolutionized computing, boosting the development of everyday electronic devices. She received the Edison Medal in 1997 and the National Medal of Science in 2009.

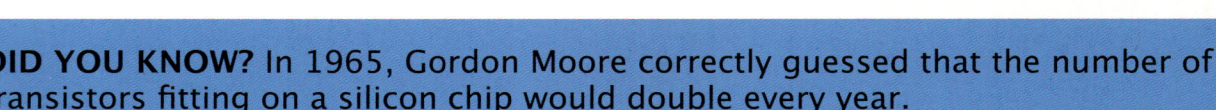

DID YOU KNOW? In 1965, Gordon Moore correctly guessed that the number of transistors fitting on a silicon chip would double every year.

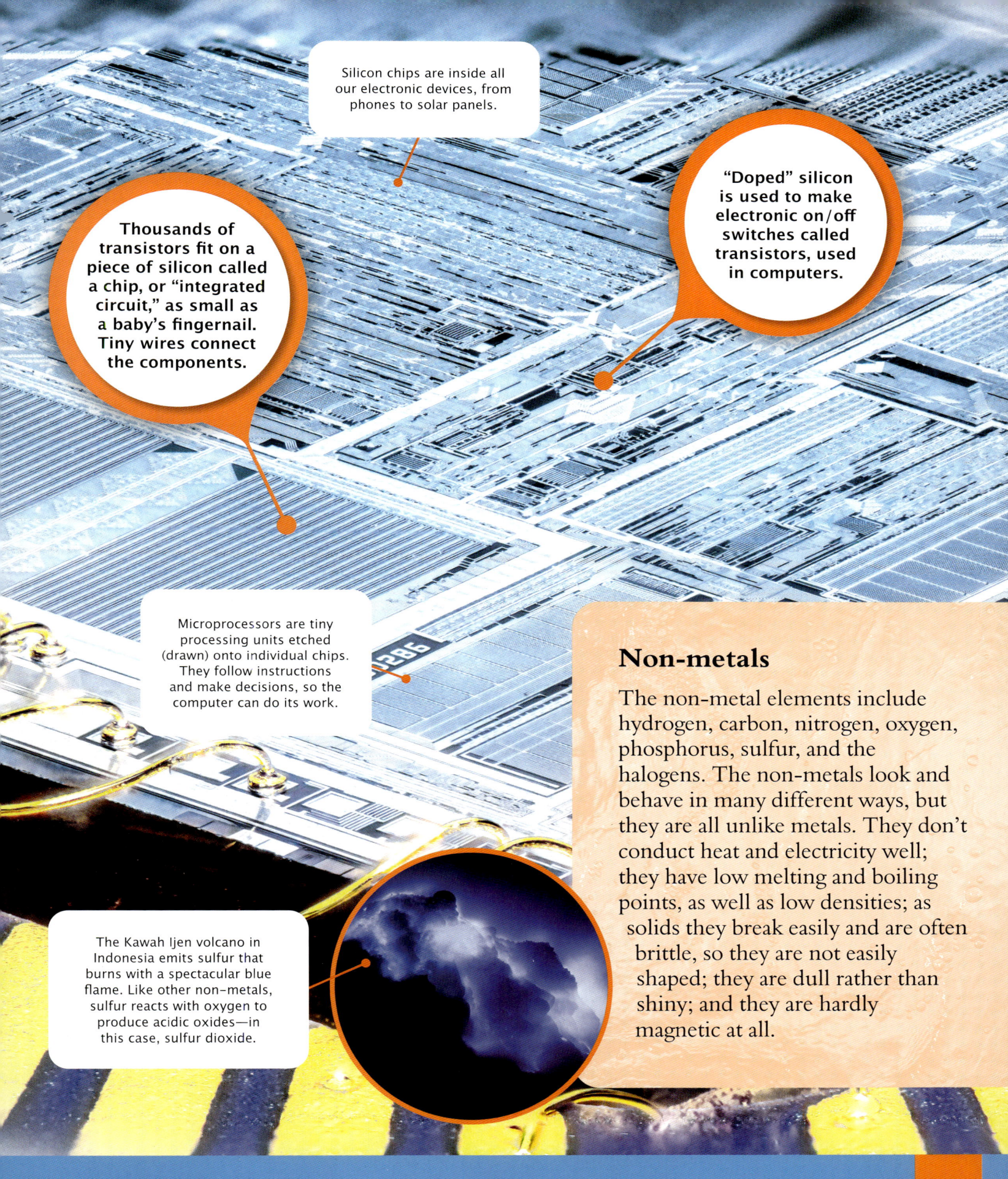

Silicon chips are inside all our electronic devices, from phones to solar panels.

Thousands of transistors fit on a piece of silicon called a chip, or "integrated circuit," as small as a baby's fingernail. Tiny wires connect the components.

"Doped" silicon is used to make electronic on/off switches called transistors, used in computers.

Microprocessors are tiny processing units etched (drawn) onto individual chips. They follow instructions and make decisions, so the computer can do its work.

The Kawah Ijen volcano in Indonesia emits sulfur that burns with a spectacular blue flame. Like other non-metals, sulfur reacts with oxygen to produce acidic oxides—in this case, sulfur dioxide.

Non-metals

The non-metal elements include hydrogen, carbon, nitrogen, oxygen, phosphorus, sulfur, and the halogens. The non-metals look and behave in many different ways, but they are all unlike metals. They don't conduct heat and electricity well; they have low melting and boiling points, as well as low densities; as solids they break easily and are often brittle, so they are not easily shaped; they are dull rather than shiny; and they are hardly magnetic at all.

Inorganic and Organic Chemicals

Atoms of carbon and hydrogen have a very important partnership. Compounds that contain carbon-hydrogen bonds are "organic," and the rest are "inorganic." Organic chemistry is all about chains of atoms—the molecules get longer as more atoms are added to the "backbone" of carbon, making series of chemicals with predictable structures and properties. The bodies of living things are made of organic chemicals that also contain other atoms, such as oxygen and nitrogen.

Hydrocarbons

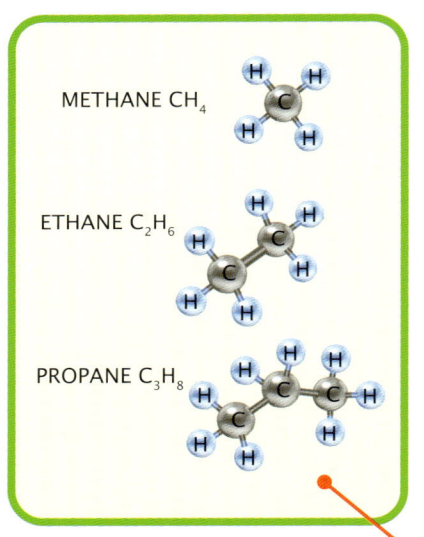

Compounds that contain only carbon and hydrogen atoms are "hydrocarbons." A carbon atom can make four bonds, and a hydrogen atom can make one bond. So, the simplest hydrocarbon is one carbon atom linked by single bonds to four hydrogen atoms—methane (CH_4). Methane is an "alkane," a hydrocarbon that contains no double bonds. Alkanes are good fuels; they burn in oxygen to produce carbon dioxide, water, and energy. They have predictable properties. For example, as alkane molecules get longer, their boiling points get higher.

Methane, ethane, and propane are the first three molecules in the alkane series. Each has one carbon and two hydrogen atoms more than the last, building up to very long chains.

HALL OF FAME: Saint Elmo Brady 1884–1966

Saint Elmo Brady was the first Black American chemist to be awarded a PhD in America. He studied carboxylic acids (molecules with a special arrangement of carbon, hydrogen, and oxygen atoms) to see how changing parts of a molecule affected its acidity. He improved ways of preparing and purifying organic acids. Brady also helped develop academic facilities at Historically Black Colleges and Universities (HBCU).

DID YOU KNOW? It's estimated that an average cow releases 375 liters (82 gallons) of methane a day from both ends—the same level of pollution as a car.

A polymer molecule is a repeating structure, with molecules of the monomer linked up like beads on a chain.

Monomers
Polymerization
Polymer

Monomers and Polymers

Polymers are long, repeating chains of many small molecules called monomers. Most (but not all) polymers are organic, and some occur naturally while others are manufactured. Plastics are manufactured polymers. The strong plastic polyvinyl chloride (PVC) is a long chain of linked vinyl chloride monomers containing carbon, hydrogen, and chlorine atoms. Natural polymers include fats, starches, proteins, and wool. Nucleic acids—the molecules that carry our genetic information—are natural polymers that are particularly large and complicated molecular structures.

Plastics are organic polymers made out of chemicals from fossil fuels. Plastic bottles are made from polymers like polyethylene terephthalate (PET). They are light, bendy, and easily shaped.

Glass bottles are made from an inorganic material, silicon dioxide (SiO_2), also known as silica—or sand. It is transparent, stiff, and brittle.

Inorganic materials (like glass and metals) and organic materials (like plastics, paper, and cardboard) all should be used responsibly— reused as much as possible and then recycled.

55

Radioactivity

Protons are the subatomic particles that identify elements, and each element has its own number of protons. But the number of neutrons is different in different isotopes (forms) of an element. An isotope with more neutrons than usual is unstable and it decays (breaks up), giving off energy as radioactive particles and rays.

Half-life

When radioisotopes (unstable isotopes) break up, their nuclei give off three types of radiation: alpha particles (made of two protons plus two neutrons), beta radiation (electrons), and electromagnetic waves (gamma rays). This changes the number of subatomic particles, which means that radioactive elements change into other elements. This radioactive decay occurs atom by atom, and it happens at different rates for different radioisotopes. A radioactive isotope's "half-life" is the time it takes for half of its nuclei to decay into the atoms of another element. The half-life of radon-222 is four days.

Nuclear radiation can harm living cells—so it can also be helpful in killing germs on food and medical equipment, or killing damaged cells in cancer patients.

Plant- and animal-based pigments in prehistoric cave paintings can be dated by carbon dating. Carbon-14 can date objects up to 60,000 years old. Other radioisotopes date older fossils.

Carbon Dating

Free neutrons—not held in an atomic nucleus—occur high in the atmosphere. When one bumps into a nitrogen atom in the air, the nitrogen gains a neutron and releases a proton—and so changes into an atom of carbon-14. Carbon-14 atoms become part of carbon dioxide molecules, and so enter the food chain through plants. Carbon-14 is weakly radioactive. All living things contain a known percentage of it, and when they die, those carbon-14 atoms decay slowly into nitrogen—the half-life is 5,700 years. How much carbon-14 is left tells us the age of long-dead things—carbon dating.

HALL OF FAME: Henri Becquerel
1852–1908

French physicist Henri Becquerel was looking for X-rays in fluorescent materials when he accidentally discovered radioactivity instead—he found that uranium sulfate granules left an image on photographic film. He received the Nobel Prize in Physics in 1903, jointly with Marie and Pierre Curie, and the becquerel unit of radioactivity was given his name.

A neutron can break up into a positive proton and an electron. The electrons released in beta radiation come from the nucleus, not from the energy shells of the atom.

Different types of radiation travel farther and are blocked by different objects—paper or skin stop alpha particles, aluminum stops beta radiation, and gamma rays are stopped by lead.

In nuclear fission reactors, neutrons bombard the isotopes of heavy elements to split the nuclei. A chain reaction releases more neutrons plus energy, which is used to make electricity.

Radioactivity disturbs the electrons of a gas in an instrument called a Geiger counter; this is measured in becquerels (Bq). The yellow and black sign is the nuclear radiation warning.

DID YOU KNOW? Earth's background radiation causes changes to genes inside cells, and so it drives natural evolution (living things developing from ancestral forms).

Chapter 3: Chemistry in Nature

The First Chemistry

It is thought that 13.8 billion years ago there was nothing but a tiny area, so full of energy that it burst in a huge, hot explosion. This is the Big Bang theory. It explains how the Universe began and how it's still expanding. The theory is supported by the elements that we can detect out in space, across the Universe.

Star Birth

Absorption spectra—the "fingerprints" of elements—show that 99.9 percent of our Sun is hydrogen and helium, with tiny amounts of other elements. Other stars are the same. Hydrogen and helium are the lightest atoms, and they formed first after the Big Bang, making gas clouds. The clouds collapsed to make protostars—balls of hot gas containing the charged particles, electrons, and hydrogen ions. Nuclear fusion began to "burn" hydrogen ions, joining the nuclei to make helium and the first starlight.

The Forging of Elements

Nuclear fusion deep in the Sun's core turns hydrogen into helium, and produces energy. As any star ages, its core runs out of hydrogen fuel and starts fusing helium instead. When helium fuel runs out, the nuclei of larger atoms start to fuse and produce the heavier elements, up to iron. When the star's core runs out of fuel, it collapses. In the case of giant stars, huge energies then produce the heaviest elements, and scatter them in a supernova explosion.

The nuclei of hydrogen isotopes deuterium and tritium have one positively charged proton (red). They fuse to make a helium nucleus with two protons. More nuclei additions make heavier elements.

DID YOU KNOW? New stars burst into life by starting nuclear fusion at temperatures above 15,000,000°C (27,000,032°F).

HALL OF FAME:
Cecilia Payne-Gaposchkin
1900–1979

British–American astronomer and astrophysicist Cecilia Payne-Gaposchkin studied at Cambridge University, UK, but, as a woman, received no degree. She became the first woman to earn an astronomy PhD from Radcliffe College, USA, and was a trailblazer for women. She saw that the Sun's spectrum had far more hydrogen and helium than other elements, and realized that hydrogen was the most abundant element in the stars.

The Horsehead Nebula is 1,600 light-years away—its light takes 1,600 years to reach Earth.

Galaxies are groups of stars. They also contain gas and dust clouds—nebulae—where stars are born. The Horsehead Nebula looks like a chess piece.

The Orion star constellation is like a dot-to-dot of a hunter with his bow. Find the star called Alnitak at the bottom of his belt to look toward the Horsehead Nebula.

The stars are chemical factories. Except for hydrogen and helium and a few others, all the elements' atoms in the Universe were made inside stars.

You are star stuff—you are made of atoms that were built in the stars billions of years ago!

Our World

Our Solar System formed 4.6 billion years ago, from a cloud of dust and gas around the Sun. Near the hot center, dust gathered into rocks, and rocks collided and clumped together, eventually forming Earth. The molten materials inside the planet arranged themselves into layers—heavier elements sank to form the core, and lighter materials floated upward, cooled, and hardened.

Young Earth

As Earth cooled, blocks of rock floating on molten rock became the crust. Volcanic gases erupted and began to form the atmosphere. When Earth had cooled to below the boiling point of water, water in the atmosphere condensed and began a rainstorm that lasted for centuries. Water gathered in hollows on the surface and made the first oceans. Today, the Earth's crust is still active and changing. It moves in huge "tectonic plates" and recycles itself through volcanic activity.

As the Moon orbits Earth, its gravity pulls on the oceans, making tides which help life thrive. Tides also drive the ocean currents, helping to keep Earth's climate stable.

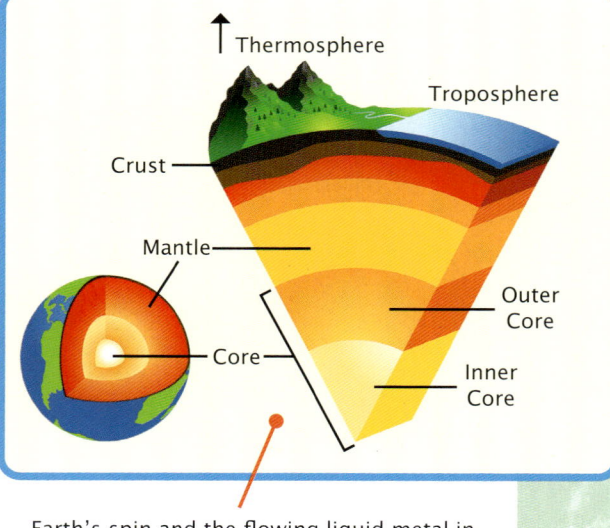

Earth's spin and the flowing liquid metal in the outer core put a magnetic field around the planet, which protects our atmosphere from solar winds and radiation from Space.

Layers of Earth

Earth has a heart of iron and nickel—a solid inner core surrounded by a liquid outer core. The layer above the core is the mantle, made of hot, slowly flowing, semi-molten rock, or magma. The crust is the outer layer of solid rock, where we live and oceans flow. Surrounding it all are the gases of the atmosphere. The atmospheric layer where we breathe air and where aircraft fly is the troposphere. The International Space Station is in the thermosphere.

DID YOU KNOW? The pale patches on the Moon's surface are anorthosite—a type of rock also found on Earth—suggesting that the Moon was once part of Earth.

On dark nights, we can see the galaxy stretching, like a highway across the sky. Our Sun is one of hundreds of billions of stars in the Milky Way galaxy.

Earth formed with just the right amounts of elements to make a rocky surface and the hot metallic core.

The young Earth had all the chemicals—such as carbon, oxygen, and hydrogen—needed to make the molecules of life, plus the Sun's energy to support life.

Earth circles the Sun in the "Goldilocks Zone"—the perfect distance where the temperature is just right for water to exist as liquid.

HALL OF FAME: Florence Bascom
1862–1945

Florence Bascom was the second US woman to earn a geology PhD, having been segregated from male classmates at Johns Hopkins University. She was the first woman to work for the US Geological Survey, and she founded the geology department of Bryn Mawr College, helping to train other women. Bascom was an expert in minerals and crystals, and she discovered that rocks that were previously thought to come from sediments from lava flows.

Rocks and Minerals

Around 150 million years ago, this rock would have been a huge sand dune. Over time, winds blew the sand into layers of different grain sizes.

Earth's crust is in the recycling business. Elements and compounds combine into minerals, which are mixed up in rocks. Over thousands of years, rocks are taken underground and back up to the surface by movements of the crust and volcanic activity.

The Rock Cycle

Rocks get broken and worn down—eroded—by forces like the weather. Rivers carry the small pieces to the sea where they are laid down as sediment. The weight of sediment layers squashes them, forming "sedimentary rocks." Rocks buried very deep, with high temperatures and pressures, change into "metamorphic rocks," such as marble. When metamorphic rocks are heated until they melt, the molten rock—magma—comes to the surface in volcanoes, and cools as "igneous rocks," such as granite.

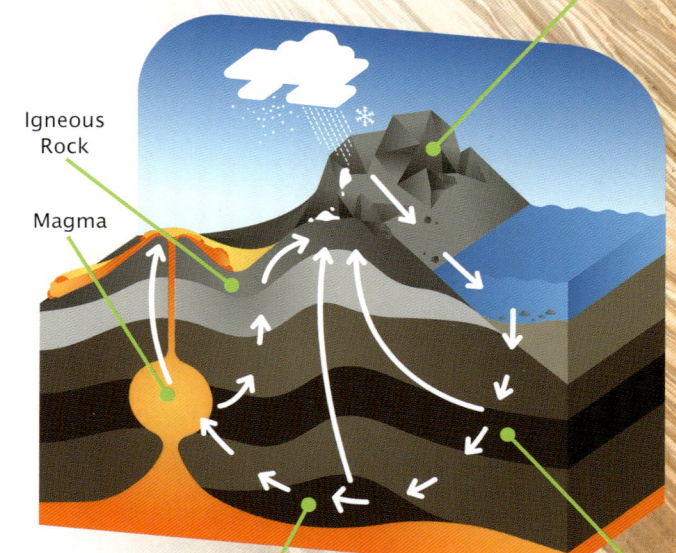

Rocks lifted to the surface are weathered and eroded, and the cycle goes on.

Igneous Rock

Magma

Metamorphic Rock

Sedimentary Rock

HALL OF FAME: Marie le Jars de Gournay
1565–1645

Marie le Jars de Gournay was the adopted daughter of the French philosopher Michel de Montaigne. As a writer and early feminist, she argued for gender equality, insisting that women had the same rights as men to education and in the workplace. She performed practical experiments using alchemy, mineralogy, and philosophy, defining herself as an educated individual in a male-dominated society.

The sandstone has been shaped by rain and wind. The weathering reveals the strata—the uncountable lines and ridges of the formation.

Rock Minerals

There are thousands of minerals, but only about 200 are common in rocks. Sedimentary rocks are often mainly one mineral. For example, limestone is made from seashells, so it's mostly calcium carbonate. It forms the metamorphic rock marble, so marble is also mainly calcium carbonate. Sandstone is made from compacted sand, which is silica, or silicon dioxide—a compound of silicon and oxygen, the two most common elements in Earth's crust. Heat and pressure turn sandstone into the metamorphic rock, quartzite.

Pumice is foamed, igneous rock. It erupted from a volcano like fizzy pop from a shaken bottle, then cooled quickly, with holes where the gas bubbles had been.

The Wave is a sandstone formation in northern Arizona, USA. The rock is unusual because it formed in the desert, not under the sea.

Eventually, the sand compacted into stone. The colors are due to iron, manganese, and other mineral salts in water seeping through the porous sandstone.

DID YOU KNOW? Earth's oldest rocks are igneous "faux amphibolites" in Canada. At 4.28 billion years old, they were probably once part of Earth's earliest crust.

Wonderful Water

Water is Earth's superpower! Our planet is just the right distance from the Sun for liquid water to exist. Without water, life on Earth could not happen, because biochemical reactions happen in water inside living cells. Luckily, Earth recycles our precious water.

Water molecules freeze into six-sided crystals. The six arms of a snow crystal are built up from a tiny, six-sided plate of frozen water vapor.

The Water Cycle

A puddle dries up, but the water isn't lost. Water evaporates—the Sun's heat constantly turns molecules at the surface of seas, lakes, and rivers into water vapor in the air. The vapor rises, cools, and condenses into droplets inside clouds. The droplets get heavier, until they fall as raindrops or snow—precipitation. Rain falling on land flows into rivers or seeps through rocks and eventually collects in the sea. Evaporation returns water to the air, and the cycle continues.

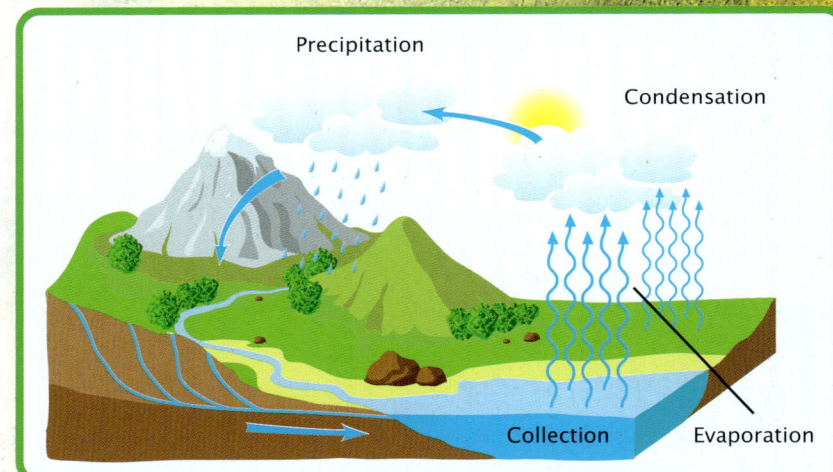

HALL OF FAME: Chandrasekhara Venkata Raman 1888–1970

Chandrasekhara Venkata Raman was an Indian physicist who won the Nobel Prize in physics in 1930, for his work in spectroscopy and the scattering of light. He described water as "the elixir of life" after standing on the edge of the desert beside the Nile valley in Egypt and seeing the difference between the empty desert sand and the fertile land by the river, where life thrived.

The River Danube splits to form a fan-shaped area of swampy land along the coastline—a delta. This river mouth in Romania is one of three main channels.

An estuary is where a river meets the sea. The fresh water of the river mingles with the seawater, so estuaries have "brackish" water that's slightly salty.

Sediments carried by the river are deposited in deltas and estuaries. The rich soils and the actions of the tides form unique ecosystems and wildlife habitats.

The River Danube begins in Germany and touches ten countries on its 2,850-km (1,770-mile) journey to the coast of the Black Sea.

Pond skaters can walk on water because water molecules stick together. This "cohesion" makes surface tension, which the insect is too light to break through.

The Very Odd Molecule

Water is a compound of hydrogen and oxygen—atoms so light that water should be a gas at room temperature, but it's a liquid. When it freezes, we'd expect it to become denser, but it expands. So solid water—ice—floats instead of sinking, which means icebergs insulate the sea underneath and help to keep Earth cool. Water molecules are "sticky," so trees can draw water up against gravity, to their topmost leaves. Water is a truly surprising substance!

DID YOU KNOW? There could be life on Mars! Water once flowed on the planet's surface, and scientists think there's still liquid water underground.

Creative Carbon

Carbon is the basis of the chemicals of life. It can make anything from a microbe to a whale because of the way its atoms bond (join up) with each other and with other atoms, especially hydrogen, oxygen, and nitrogen. The carbon acts like a backbone, stringing all these atoms together into long, strong molecules that make the carbohydrates, proteins, and fats that build living things.

Chemical Factories

Plants are natural factories. They absorb carbon dioxide (a compound of carbon and oxygen) from the air and use it, with water, to make glucose (a sugar) and oxygen. This is photosynthesis. In turn, the glucose molecules make bigger carbohydrates that build the plant's body.

You are a chemical factory, too! When you eat a plant or an animal in the food chain, you use its carbon and other atoms to make proteins that build your own body. This happens in nearly all your body cells—the chemical instructions are kept in long molecules called DNA, packed into every cell.

Photosynthesis makes glucose, a molecule with six oxygen and 12 hydrogen atoms on a chain of six carbon atoms. Glucose makes bigger molecules that build living things.

Coal, gas, and oil are the remains of plants and animals that died millions of years ago. Burning these fossil fuels releases carbon dioxide into the atmosphere.

The Carbon Cycle

Living things return their carbon atoms to the environment in different ways. Carbon dioxide is made during respiration (a process that releases energy), and it leaves your body when you breathe out. Carbon is also given out in waste, like fallen leaves or our poop, and the bodies of dead plants and animals. Organisms in the soil, called decomposers, help this waste break down into simple chemicals that plants can absorb. The decomposers release carbon dioxide into the air during respiration. Plants take the carbon dioxide to make sugars, and the carbon cycle goes on.

HALL OF FAME: Jan Baptista van Helmont 1580–1644

In 1634, Belgian chemist Jan Baptista van Helmont planted a young willow tree in a pot of soil. After five years, the tree was 30 times heavier, but the soil's weight had hardly changed. This experiment showed that the plant was getting nutrients from somewhere other than the soil, and it helped later scientists to discover photosynthesis.

Think of trees and people as a team! Trees make oxygen for us to breathe, and we make carbon dioxide for them to use in photosynthesis.

When trees photosynthesize, they lock carbon into their bodies. This helps reduce carbon dioxide in the atmosphere, and that helps control global warming.

Plants contain chlorophyll, a green pigment that absorbs energy from sunlight. They use the energy to power photosynthesis.

DID YOU KNOW? More than half of your body weight is water—and if you take away the water, half of what's left is carbon!

Essential Oxygen

Many living things need oxygen for respiration—the process of getting energy from glucose. Oxygen gas makes up around 21 percent of Earth's atmosphere. Like other chemicals widely used in biological processes, it's never used up, but keeps being recycled through the environment.

The Element

Oxygen is the third most abundant element in the Universe, after hydrogen and helium. It's the most common element in Earth's crust, making up 47 percent, mostly combined with silicon. It's also the most common element in your body, mainly inside water. Pure oxygen has no color, smell, or taste. It reacts readily with other elements to make compounds called oxides. Oxides are everywhere—water is an oxide of hydrogen, sand is an oxide of silicon, and rust is an oxide of iron.

Many land animals breathe air and oxygen into their lungs, but a fish gets oxygen by taking water into its mouth and passing it over feathery gills in its head.

Oxidation is the addition of oxygen atoms to chemicals. An apple goes brown after it's cut and exposed to air, because of an oxidation reaction.

The Oxygen Cycle

Plants on land and sea drive the oxygen cycle by photosynthesis—they use the Sun's energy to turn carbon dioxide and water into glucose and oxygen. The oxygen is released to the air, and plants and animals use it for respiration. Respiration is the opposite of photosynthesis—it turns glucose and oxygen into carbon dioxide and water. At the same time, energy is released which the animal or plant uses for chemical reactions in its cells.

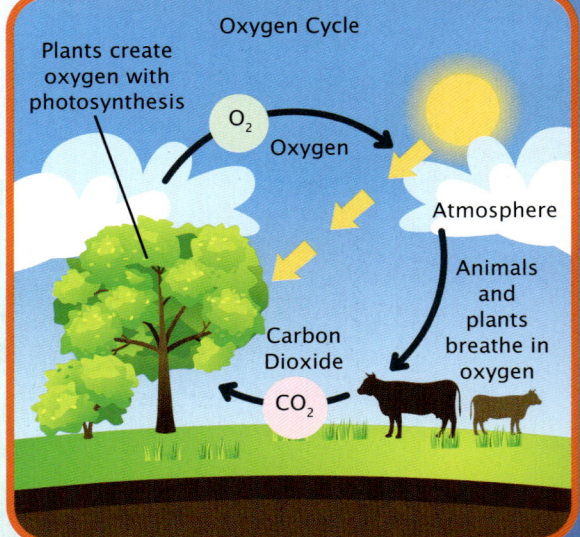

Oxygen Cycle
- Plants create oxygen with photosynthesis
- Oxygen (O_2)
- Atmosphere
- Animals and plants breathe in oxygen
- Carbon Dioxide (CO_2)

DID YOU KNOW? Giant dragonflies lived 300 million years ago, when higher oxygen levels meant that their tiny breathing tubes could get more oxygen.

Oxygen moves from the seawater into blood vessels in the gills, and it's carried by the blood to the fish's body cells, where respiration happens.

Tiny sea-living organisms called cyanobacteria started to produce oxygen by photosynthesis about 2.5 billion years ago, and so began to add oxygen to Earth's atmosphere.

Respiration with oxygen is "aerobic." "Anaerobic" respiration—without oxygen—can also happen, for example, in our muscle cells during heavy exercise.

HALL OF FAME: Joseph Priestley
1733–1804

Carl Wilhelm Scheele produced oxygen in Sweden in 1771, but did not publish his discovery until six years later. Meanwhile, English scientist Joseph Priestley published his discovery of oxygen in 1774, after collecting the gas that was produced from mercuric oxide heated by the Sun. Priestley found that the gas made breathing easier and made a candle burn more brightly.

Team Nitrogen

Nitrogen makes up most of Earth's atmosphere. Living things need it to make proteins, but there's a problem. Nitrogen atoms in the air bond strongly in pairs, and the bonds must be broken before the nitrogen atoms can form other compounds. Luckily a team of soil bacteria and plants called legumes (peas and clover) can help.

Nitrogen atoms can make three bonds, so two atoms link up with very strong triple bonds, making an unreactive molecule of the diatomic element N_2.

The Nitrogen Cycle

Helped by plants like clover, "nitrogen-fixing" bacteria recycle nitrogen in the environment. The bacteria "fix" nitrogen—they separate the N_2 atoms so they can combine in compounds called nitrates. The nitrates are absorbed by plants and used to make proteins. Nitrogen atoms pass through the food chain until they return to the soil in plant and animal waste or dead bodies. Then different bacteria recycle them into either nitrates in the soil or pure nitrogen in the air.

Lightning, volcanoes, and fire also "fix" nitrogen by breaking apart the N_2 molecules, allowing the free nitrogen atoms to combine with other elements.

Haber Process

The Haber process is an industrial method that fixes nitrogen by turning nitrogen gas and hydrogen into ammonia (NH_3). The ammonia is made into nitrogen-based fertilizers like ammonium nitrate, which help farm crops grow. However, overuse of fertilizers can put too much nitrate into rivers, and upset the nitrogen cycle. Alternative methods of adding nitrates to soils are crop rotation with legumes, and the use of natural fertilizers like manure (poop).

DID YOU KNOW? Early Egyptian alchemists made "sal ammonia" (ammonium chloride) for smelling salts by heating dung (poop) and urine (pee) with salt.

HALL OF FAME:
Samuel Massie
1919–2005

Distinguished chemist Samuel Massie went to Agricultural Mechanical Normal College after being prevented from going to the University of Arkansas, USA, because of his race. His research included developing agents to protect soldiers from poisonous gases, investigating how pollution from ships affected sea life, and studying nitrogen and sulfur compounds for treating infectious diseases.

Some nitrates in the soil are converted back into N_2 in the air by "denitrifying bacteria."

Farmers enrich their fields by planting fields of clover. Nitrogen-fixing bacteria in the clover roots turn nitrogen from the air into nitrates.

The bacteria form nodules on the clover roots, where nitrates build up. The clover uses the nitrates, and animals that eat the clover gain the nitrogen.

Decomposers—tiny organisms in the soil—and "nitrifying bacteria" break down the waste from living and dead animals and plants, making more nitrates available.

Crucial Glucose

Every living thing is building up and breaking down molecules all the time. All these chemical reactions together are its metabolism. Glucose is particularly important in metabolism. It's a monosaccharide—a simple sugar made of just one molecule.

Getting Glucose

Plants make glucose from carbon dioxide and water, with energy from the Sun. They use the glucose to make bigger molecules, like cellulose and starch. Plants are the producers in the food chain, and we are the consumers—we get glucose from plants when we eat starchy foods like bread, rice, and potatoes. Digestion breaks the starches down into simple sugars, which our blood carries to our tissues. There, tiny cell structures release energy from the molecules, for use in cell processes.

> Green parts of plants absorb sunlight and trap the energy in glucose molecules during photosynthesis. Energy is stored in the bonds between atoms inside the molecules.

> The formula of a glucose molecule is $C_6H_{12}O_6$ because it has six carbon atoms, with 12 hydrogen atoms and six oxygen atoms attached.

Making Polysaccharides

Monosaccharides join up in chains to make long natural polymers called polysaccharides. Glucose molecules ($C_6H_{12}O_6$) link up to make polysaccharides, like starch and cellulose. As each glucose molecule links up, a water molecule (H_2O) is lost, so the polysaccharide formula is $(C_6H_{10}O_5)n$, where n means any number of repeated molecules. Starch is an untidy molecule, used to store energy. Cellulose is a straight molecule, used to build strong structures like tree trunks.

DID YOU KNOW? The human brain accounts for around 2 percent of body weight, but uses around 20 percent of the glucose energy needed by the body.

Plants can't live around deep-sea hydrothermal vents with no light. Instead, the food chain producers are bacteria, making glucose from hydrogen sulfide and methane by chemosynthesis.

Glucose is stored as starch in leaves, stems, roots, and seeds—or in fruit, like pumpkins. The energy-giving, digestible nutrients we get from starchy foods are carbohydrates.

Glucose is carried from the leaves to the plant's cells. It is used either to release energy during respiration, or stored, or used to make larger molecules to grow the plant's body.

HALL OF FAME: Marie Maynard Daly
1921–2003

Marie Maynard Daly was the first Black woman to earn a PhD in chemistry in the United States. Her studies focused on the role of enzymes in starch digestion, and on the structure and biochemical activities of the cell nucleus. Daly taught biochemistry and became a professor at the Albert Einstein College of Medicine. She also pushed to enroll more minority students in medical and scientific studies.

Plant Chemicals

Plants and fungi don't seem to do much, but in fact they're busy doing amazing chemistry! There are chemical reactions going on in their bodies all the time—their metabolism. A plant has the important "job" of trapping energy and making glucose to build its body—but the chemicals of photosynthesis aren't the only ones involved in its everyday life.

Many plants contain poisonous chemicals that taste bad, to discourage munching. Ragwort (Senecio) is a British wildflower that contains toxic chemicals called alkaloids.

Plant Hormones

Plants need light, and they can seek it out. They have hormones (chemical messengers) called auxins that control the growth of their root and shoot tips. Auxin in a shoot tip moves away from light. It concentrates in the shadiest side of the shoot and makes the cells there grow faster—so the shoot bends toward the light. Auxin in a root tip diffuses downward in response to gravity. It makes the cells grow slower on the underside of the tip, so the root grows down.

Most of a fungus's body is hidden underground. Fungi aren't plants, so they don't photosynthesize. They break down decaying material to get to the nutrients inside.

Wood Wide Web

Trees look solitary, but they share food and chemical messages as a community. Hyphae—threads of fungi—link the tree roots underground, making a network called the "Wood Wide Web." The fungi get glucose from the trees and give nutrients in return, and they make a pathway for trees to share stuff. Older trees feed sugars to seedlings, while less friendly trees send harmful chemicals. Plants also know when other plants close to them are being eaten, and respond by making protective chemicals in their own leaves.

DID YOU KNOW? The Venus flytrap can count! When an insect touches its sensory hairs, the plant counts two touches before snapping the trap shut.

HALL OF FAME: Salimuzzaman Siddiqui
1897–1994

Salimuzzaman Siddiqui was an Indian-born scientist who earned his PhD in organic chemistry in Germany in 1927, and studied traditional herbal therapies. He isolated Indian snakeroot (Rauvolfia) alkaloids, used as sedatives and in treating high blood pressure, and he extracted compounds from neem tree oil to treat infections. He was Professor of Chemistry at Karachi University, and did much to advance developments in science in Pakistan.

The black and yellow markings of this type of caterpillar, and the red and black markings of the adult moth, warn predators that they taste bad—so they are less likely to get eaten!

Cinnabar moth caterpillars feed on ragwort. The plant's toxins don't hurt them, but stay in the caterpillars' bodies, making them poisonous too.

Ragwort can harm farm animals if they eat a lot of it, but it supplies nectar and pollen to many insects, and it's important for biodiversity.

Tree leaves contain red and yellow pigments, which act as sunscreen. They're hidden by the green pigment chlorophyll, so we see them in the fall, when trees stop producing chlorophyll.

Body Chemistry

Out of all the millions of chemicals, just a few organic (carbon-based) compounds are used by living organisms. These "biomolecules" include proteins, carbohydrates, lipids, and nucleic acids. Animal bodies are built from these molecules, and the chemical reactions that involve them are their metabolism.

Proteins

Proteins are polymers (long-chain molecules) made of building blocks called amino acids. There are around 20,000 different proteins in your body, including hemoglobin, which carries oxygen in red blood cells. Your muscles work because proteins contract and slide over each other. Enzymes that break down your food are proteins, and so are hormones—chemical messengers like adrenaline and insulin. Nucleic acids, found in chromosomes, are not proteins, but they are the chemicals that carry the code to make proteins.

Social insects like ants live and work together in a colony. They communicate with chemical signals—pheromones—that carry messages to all the members of the colony.

Skeletons

Skeletons are hard structures that protect animals' soft tissues, hold them up, and help them move. Animals with backbones have a bony, inner skeleton made of a protein called collagen, strengthened with calcium phosphate. Many invertebrates have skeletons outside their bodies—an insect's exoskeleton is like a suit of armor made of a polymer called chitin, while snails and other mollusks build shells out of calcium carbonate.

Jellyfish are 95 percent water! Even their skeletons are made of water. The water pressure gives their bodies shape, and they squirt it out one way to move in the opposite direction.

DID YOU KNOW? Your smile shows off the hardest compound in your body—tooth enamel. It's made of a calcium phosphate crystal called hydroxyapatite.

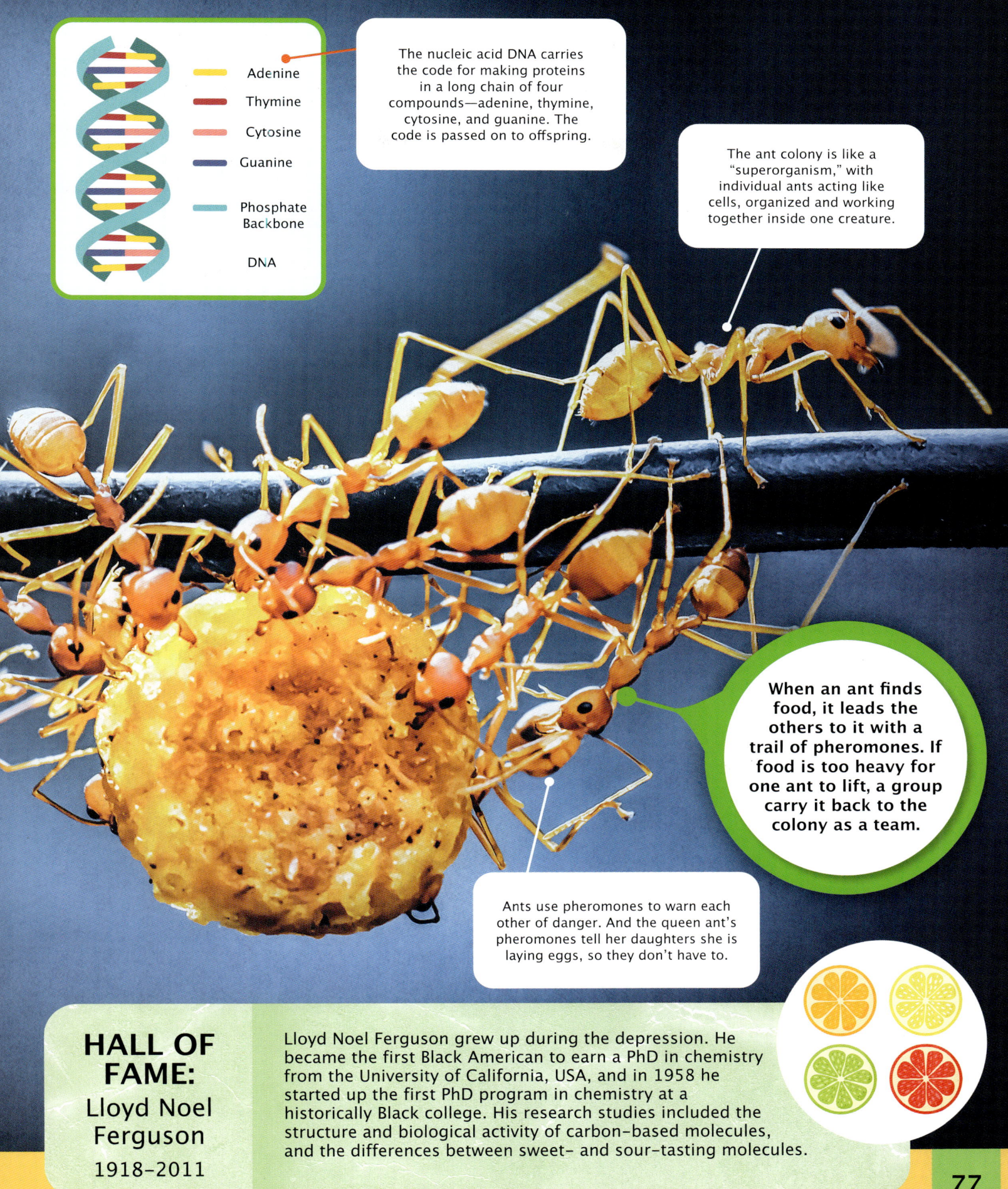

PHYSICS

Science is all about learning and understanding the rules that make the Universe and everything in it work. At the heart of this is physics, the science of matter, the stuff from which everything is made, and the energy that moves and transforms that matter. Understanding physics allows us to explain how everything else works.

Forces and Matter

Forces are some of the fundamentals of physics. From the pull of gravity to the push of pressure, from floating to sinking, physics is there. Physics sets out laws of motion that explain how things move. These laws look at the mass of an object—a measure of how much matter is inside it—and the size and direction of the forces pushing on it.

Energy and Motion

Physicists define energy as the ability to do work. Energy can come in many different forms, such as the energy of movement, heat, or sound. It can be changed, or transformed, from one form into another, but it can never be created or destroyed. Everything that happens in the Universe, from a bouncing ball to an exploding star, requires energy.

These acrobats are expert movers. They are relying on physics to put on their show.

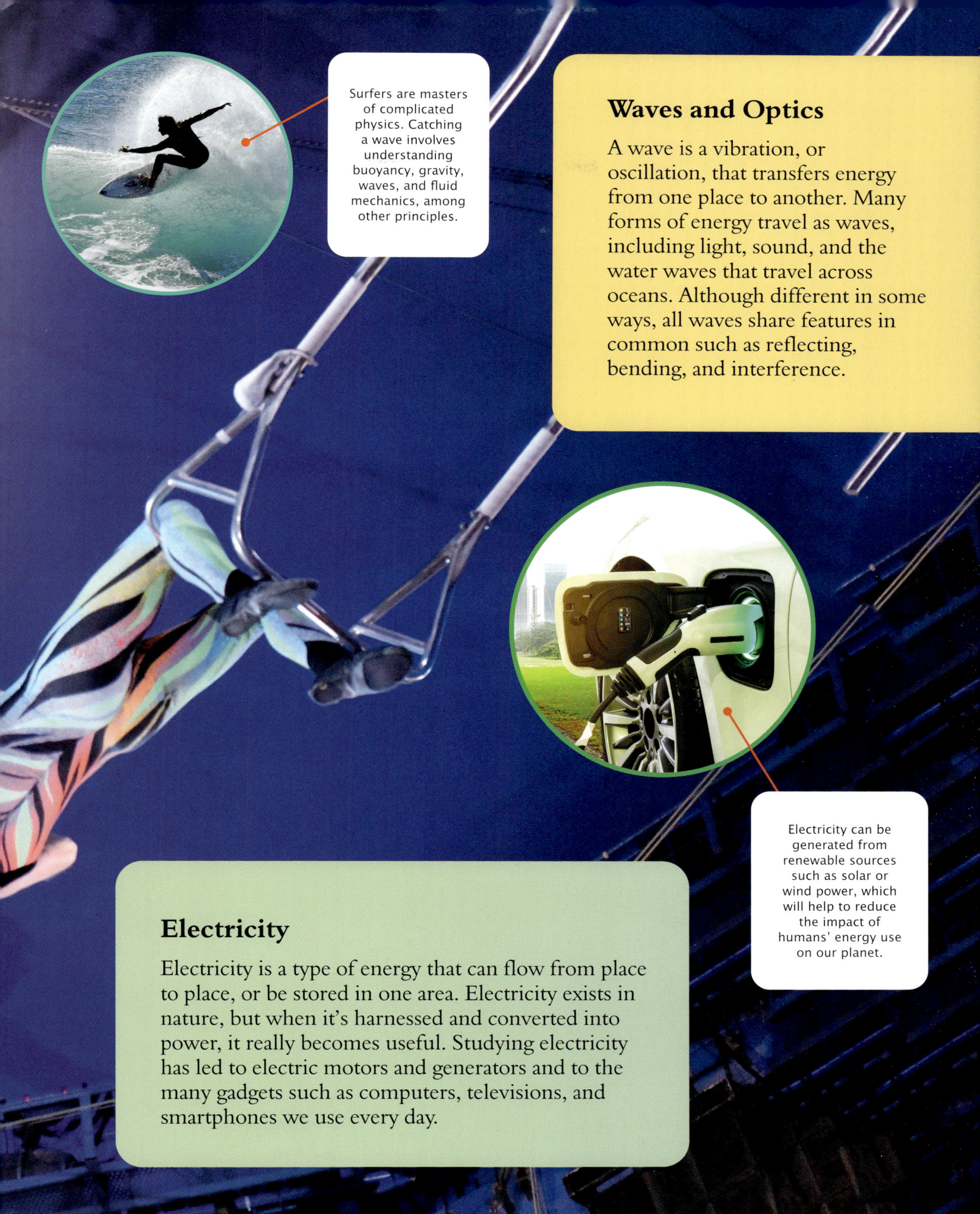

Surfers are masters of complicated physics. Catching a wave involves understanding buoyancy, gravity, waves, and fluid mechanics, among other principles.

Waves and Optics

A wave is a vibration, or oscillation, that transfers energy from one place to another. Many forms of energy travel as waves, including light, sound, and the water waves that travel across oceans. Although different in some ways, all waves share features in common such as reflecting, bending, and interference.

Electricity can be generated from renewable sources such as solar or wind power, which will help to reduce the impact of humans' energy use on our planet.

Electricity

Electricity is a type of energy that can flow from place to place, or be stored in one area. Electricity exists in nature, but when it's harnessed and converted into power, it really becomes useful. Studying electricity has led to electric motors and generators and to the many gadgets such as computers, televisions, and smartphones we use every day.

Chapter 4: Forces and Matter

Matter and Energy

Matter is anything that takes up space and has mass. All matter is made of tiny particles called atoms. Your body, your home, and the trees, oceans, rocks, and everything else on planet Earth are all made of matter. There are different kinds of energy, but they are not types of matter. They don't take up space or have mass.

Although we cannot see it, we are surrounded by matter all the time. These are the gases in the air!

States of Matter

Matter exists in three main states—solid, liquid, and gas. The state is the result of how atoms can move within the matter. Matter changes state as it becomes hotter or colder. In a cold solid, the atoms are arranged into a fixed shape and volume. Solids melt into liquids as they warm up. Liquids have a fixed volume but no fixed shape. Heating a liquid until it boils makes a gas, which has neither a fixed shape nor volume.

The most familiar matter that changes between solid, liquid, and gas is water. On Earth, it is solid as ice, liquid as water, and gas as steam.

Matter and Energy

When matter is heated, the heat energy makes the atoms vibrate. As matter cools—loses heat—the atoms move less. A change of state is a physical change, because the matter remains the same substance. Atoms in it remain joined in the same way but are in different places. Matter can also lose or gain energy in a chemical change. A chemical change rearranges the atoms into different substances. For example, burning a log changes it from wood to smoke, gas, and ash.

An explosion is a rapid chemical change that changes the nature of the matter. Some is converted to smoke or gases. A lot of energy is released as heat.

80 **DID YOU KNOW?** There is a fourth state of matter called plasma. It can be created when a gas gets very hot, or is affected by electricity. The Sun is made from plasma. In fact, 99 percent of all known matter in the Universe is plasma!

HALL OF FAME:
John Dalton
1766–1844

The idea of matter being made of atoms dates back to ancient Greece. John Dalton, a scientist from Britain, introduced modern atomic theory in 1801. He suggested that atoms of different types combine to make different substances. They always combine in the same proportions to make the same new substance.

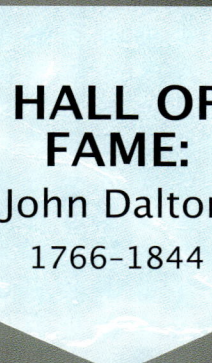

Matter can be pushed and pulled on by forces. The forces can make the matter move and change shape.

Matter exists as millions of different substances. Each substance has its own properties, such as color, density, and stretchiness.

Electromagnetism

Electromagnetism is one of the basic forces in the universe. It produces electricity, magnetism, and light, and it holds matter together. All particles have an electric charge, which can be positive, negative, or neutral. Objects with opposite charges are attracted to each other, while those with like charges push each other away. The charge comes from subatomic particles, most often the negatively charged electron. Anything that gains extra electrons has a negative charge, while anything that loses them has a positive charge.

Light Radiation

Visible light, radio waves, infrared, and X-rays are all forms of electromagnetic radiation. When electrons in an atom give out energy, they release a burst of light or other radiation. When that radiation hits another atom, it might be taken in by an electron (absorbed) or bounce off again (reflected).

A laser is an instrument that can produce a powerful beam of light. The word laser stands for the scientific term that explains how a laser beam is produced: Light Amplification by the Stimulated Emission of Radiation."

Electric Currents

Electricity is a flow of charged particles—most often electrons. The electrons move from an area where there are many electrons to a place where there are fewer. Batteries and other sources of electricity keep the current flowing by always adding more electrons at one end and removing them from the other.

Batteries use chemical reactions to produce electric currents.

DID YOU KNOW? The force of electromagnetism is 137 times weaker than the strong force that holds the particles of the atomic nucleus together, but many trillions of times stronger than the force of gravity.

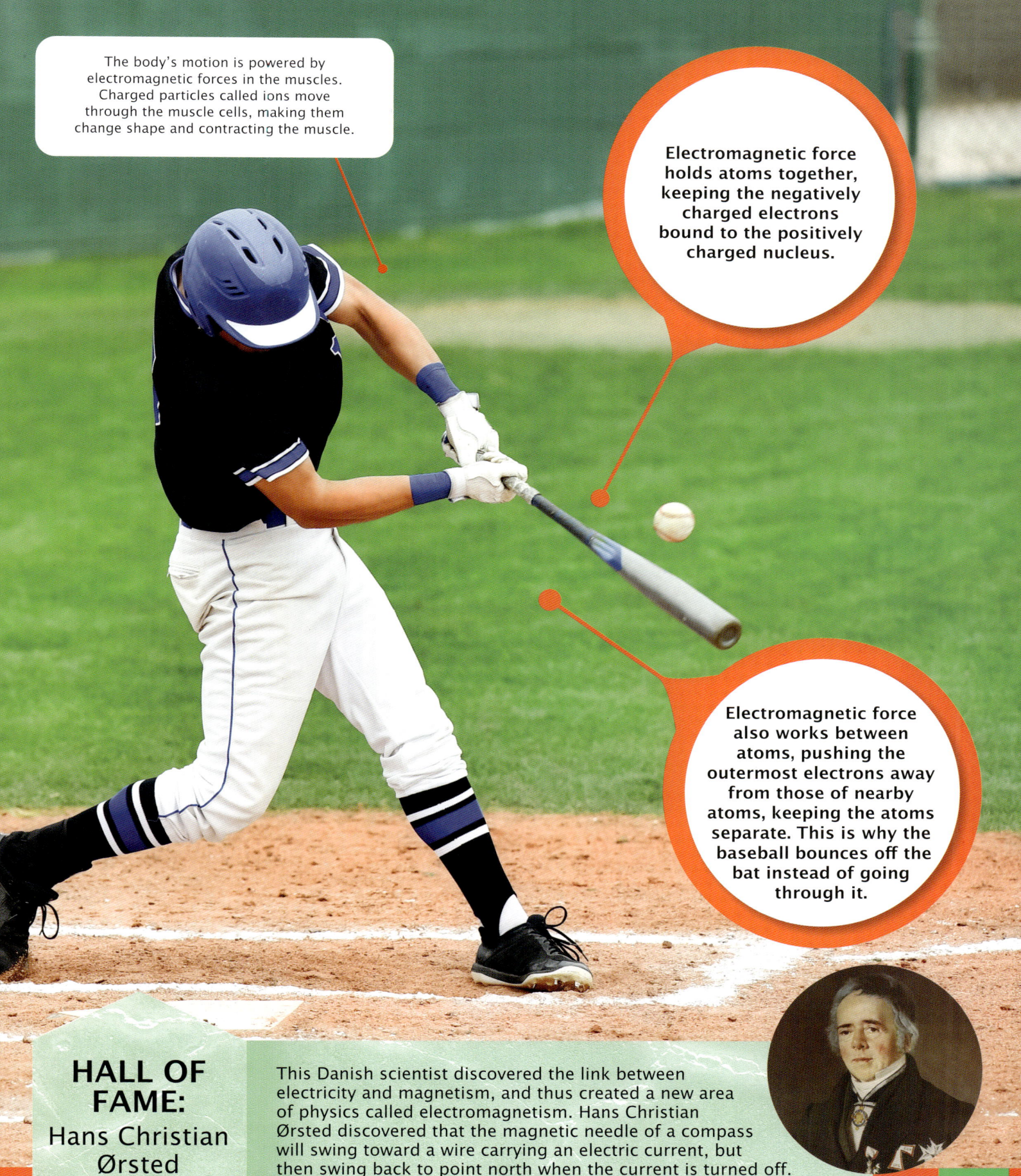

The body's motion is powered by electromagnetic forces in the muscles. Charged particles called ions move through the muscle cells, making them change shape and contracting the muscle.

Electromagnetic force holds atoms together, keeping the negatively charged electrons bound to the positively charged nucleus.

Electromagnetic force also works between atoms, pushing the outermost electrons away from those of nearby atoms, keeping the atoms separate. This is why the baseball bounces off the bat instead of going through it.

HALL OF FAME:
Hans Christian Ørsted
1777–1851

This Danish scientist discovered the link between electricity and magnetism, and thus created a new area of physics called electromagnetism. Hans Christian Ørsted discovered that the magnetic needle of a compass will swing toward a wire carrying an electric current, but then swing back to point north when the current is turned off. In 1820 he published his finding that an electric current produces a magnetic field around the wire it travels through.

Magnets

Magnetism is a complex process based on how atoms are aligned inside a material. It is most obvious in metals like iron and nickel. A magnet produces a magnetic field around itself that affects how particles or objects will line up or arrange themselves. All magnets have a north pole and a south pole. Opposite poles attract each other while like poles repel each other.

Electromagnets

There are two main types of magnet—permanent magnets and electromagnets. A permanent magnet is always magnetic. An electromagnet can have its magnetic force turned on and off. It is usually made from iron with a copper coil around it. An electric current in the coil turns the iron into a magnet, but only while the current flows. Electromagnets are very useful. For example, they can be used as parts in a machine that move when the electric current is turned on.

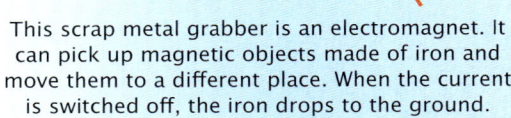

This scrap metal grabber is an electromagnet. It can pick up magnetic objects made of iron and move them to a different place. When the current is switched off, the iron drops to the ground.

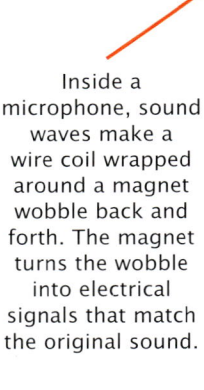

Inside a microphone, sound waves make a wire coil wrapped around a magnet wobble back and forth. The magnet turns the wobble into electrical signals that match the original sound.

Used in Gadgets

Magnets are important parts of electrical devices. Very early computers—from long before microchips were invented—used electromagnetic switches to connect and disconnect their circuits. Magnetic storage devices, from hard disk drives to the magnetic strips on credit cards, store data as tiny positive and negative charges on a surface, coded as 0s and 1s. These are written to and read from the surface by an electromagnet.

DID YOU KNOW? Earth's magnetic field is created by the currents flowing through the planet's hot liquid metal core.

Magnetic forces are strongest at the poles of a magnet. There are two poles on each magnet, north and south. Like poles repel each other, and opposite poles attract each other.

A magnet is surrounded by a force field that runs from one pole to the other. Magnetic objects that enter the field are pulled toward the magnet.

Magnets have most effect on ferromagnetic materials, which include the metals iron, nickel, and cobalt. Other metals are only weakly affected by magnets, and non-metals like plastic are not affected at all.

HALL OF FAME: William Gilbert 1544-1603

William Gilbert was a British doctor who was once in charge of looking after the health Queen Elizabeth I, but he also carried out experiments to test his theory that Earth is a giant magnet. He carved spheres out of lodestone, a naturally magnetic, iron-rich rock. He then placed a compass on different parts of each sphere. The magnetic needle always pointed to the sphere's north pole, just as a magnetic compass points to Earth's North Pole.

Gravity

What goes up must come down—because of the force of gravity. Gravity is a force of attraction between all objects that have mass. More massive bodies produce a stronger gravitational pull than less massive ones. The force between two objects also becomes stronger as the objects get closer together.

> The force of Earth's gravity makes objects—including these skydivers—accelerate toward the middle of the planet.

Planetary Orbits

Gravity is the force that keeps a planet in orbit around a star, and moons in orbit around a planet. The gravity of the larger body—the star, for example—is pulling on the smaller one. The planet does not fall into the star because it is traveling too fast. It is falling all the time, but the direction that would take it to the star's middle (or "down") keeps changing, so it constantly "falls" around!

Jupiter has several dozen moons that are all held in place by gravity.

Measuring Gravity

The strength of gravity acting between bodies depends on how much mass they each have. The link between a mass and the gravity it produces is a number called the gravitational constant (G), sometimes referred to as "big G." G is the same all over the Universe and can be used to calculate the pull of gravity between objects anywhere.

In the 1790s, Henry Cavendish measured the pull of gravity between large and small balls so he could figure out a value for G, then used the value to calculate the weight of Earth.

HALL OF FAME: Isaac Newton
1642–1727

The law of gravity was described by Isaac Newton in the mid-seventeenth century. He had returned to his home in the country to escape an epidemic of plague. He sat in his garden and saw an apple fall from a tree to the ground. At that moment, Newton understood how gravity acts between two bodies. Newton also explained the laws of motion, studied light, and created the reflecting telescope, improving on the refracting telescopes then available.

The initial downward pull of gravity is soon balanced by the upward push of resistance from the air. The skydivers then stop accelerating, but keep falling at a constant speed called the terminal velocity.

Gravity is a two-way force, so as well as the skydivers falling to Earth, Earth is being pulled up toward the skydivers. However, the planet is so much bigger that it moves only a very tiny amount—the skydivers fall much farther!

DID YOU KNOW? The gravitational effect of a black hole is so powerful that even light cannot escape it—and that is why it is a black hole.

Weight and Mass

The words "weight" and "mass" are often used interchangeably, but they have different meanings. Mass is a measure of how much matter is in an object. That object's weight is the force of gravity pulling on it. On Earth, the two measures are the same, but on the Moon, where gravity is lower, an object would weigh less than on Earth. The object's mass would always be the same, however.

The mass of these weights is a fixed amount, a measure of the amount of matter in them. That never changes—even in space.

Making Measurements

Weight and mass are measured on Earth using scales that measure the force pulling down on an object. However, mass is also a measure of how an object will resist moving. When floating in space, an object will not push down on scales at all and so it is "weightless," but its mass means it still needs a big push to get it moving—and to stop it again.

On electronic scales, the weight is measured by how much an object presses down on a pad inside.

HALL OF FAME: Andrea Ghez
1965-

Andrea Ghez is an American astronomer who discovered the most massive thing in the galaxy. In 2012, Ghez showed that there is a black hole in the middle of the Milky Way, called Sagittarius A*. Ghez used big telescopes to watch how the gravity from the black hole made nearby stars move very fast. She used those speeds to calculate the pull of gravity from the black hole, which told her that Sagittarius A* had a mass 4 million times that of the Sun!

DID YOU KNOW? The pilots of fast fighter jets experience G-force as they make tight turns. A G-force of 2 is twice the natural pull of Earth's gravity, so the pilot's body weighs twice as much.

An astronaut in space is weightless but not massless.

Weightless in Space

Astronauts on the International Space Station have no weight, since they are in constant freefall, but they need to monitor their mass to make sure they stay healthy. This is measured using a special device that calculates how much force is needed to move their body against a spring of known resistance. An object with mass still needs force to move it, even when it has no weight.

The weight of this bar depends on the force of gravity pulling it down to the ground. On Jupiter, where the gravity is stronger, its weight would be nearly three times as much!

The weightlifter has to create a force greater than gravity to get the weight off the ground and above her head.

Friction and Drag

Matter is seldom smooth, and is always rubbing against other matter—even at the atomic level. This rubbing creates resistance forces called friction and drag. Friction occurs when two solids resist moving against each other, while drag is the resistance of an object moving through a liquid or gas.

Friction is reduced by lubricants. These are usually liquids that form a slippery layer between the solid surfaces. This reduces how much the surfaces rub, and so means there is less friction.

Air Resistance

When an aircraft moves through the air, the air rushing around it works against the force moving the craft forward. This creates drag, or air resistance, that slows it. When an object is falling, drag slows its fall. The size of the drag force depends on the surface area of the object. A wide but light object like a feather flutters through the air, but a heavy and pointed object like a spear slices through it with little drag.

A parachute has a very large surface area, which creates a large drag force. The drag reduces the speed of the fall so a person can land safely.

Rough Surfaces

Friction is produced when solid objects slide past each other. The rough surfaces will snag on each other and hamper the sliding motion. Friction is present in all moving parts and is why machines always grind to halt unless they are given another push.

The treads on a tractor wheel are designed to create a large friction force so they grab the ground even in slippery conditions.

HALL OF FAME: Agnès Poulbot
1967–

This French engineer is one of the world's leading designers of car tires. With her colleague, Jacques Barraud, she designed tires with layers of shallow tread that grip the road and channel water away from the wheel. As one layer wears away, a new layer is revealed, restoring the tire's grip. Good tires mean the car uses less energy to move forward, so it is more fuel-efficient.

Ice skates have a thin blade, so there is little contact between the skate and the ice. The blade and ice are both smooth, and a thin layer of liquid water lubricates the movement, making it almost frictionless.

Each liquid has a particular "runniness" or viscosity. Water has a low viscosity so it produces less drag and friction. Honey is a liquid with a much higher viscosity and so it would be much harder to skate over!

DID YOU KNOW? Shooting stars are little rocks from space that burn up in the high atmosphere. They burn because the strong air resistance makes them very hot.

Pressure

Pressure measures the amount of force working in a specific area. In an area of high pressure, a large force is working in a small area. The same force acting in a wider area makes a lower pressure.

Atmospheric Pressure

The air around us is not weightless. Instead, it is pushing down on the ground (and us) all the time. This is measured as atmospheric pressure (also known as air pressure). Standard air pressure is equivalent to the weight of several large pitchers of water pushing onto every inch of your skin. Your body is used to this pressure—it's normal.

The air in a car tire is around 2.5 times higher than the normal air pressure. That high-pressure air inside makes the tire stiff but still allows it to bend when needed.

Air pressure is linked to weather changes. When the pressure drops, a storm is coming.

Scuba divers can only go to about 40 m (130 ft) underwater. After that, the pressure pushing on the body makes it too hard for air from the tanks to get into a diver's lungs.

Water Pressure

Water is denser than air, so the pressure from water is much greater. At the ocean's surface a swimmer experiences normal air pressure. If they swim down 10 m (33 ft), that pressure will have doubled as the weight of the water pushes on the body. In the deepest parts of the ocean, the pressure is so high that it would squash the body completely. Only the toughest submersibles can visit there.

DID YOU KNOW? Air pressure is lower at higher altitudes because the air is less dense. At the top of Mount Everest the air pressure is only one-third of the pressure at sea level.

A pump is a machine that pushes liquids or gases. The air pump used here keeps pushing air into the tire to increase the pressure.

A one-way valve allows the air to flow into the tire, but stops it from blowing out again.

HALL OF FAME:
Blaise Pascal
1623–1662

This French mathematician and physicist made the first measurements of air pressure and explained how it worked. As a result, the scientific unit of pressure is the pascal (Pa). Normal air pressure at sea level is 100,000 Pa. Blaise Pascal also helped create the mathematics of probability, which uses numbers to figure out how likely it is that something will happen.

Dark Matter

Astronomers have measured all of the matter in the Universe that they can see—all of the stars and galaxies—but they have found that the mass of the Universe is much greater than the total mass of all of the stars and galaxies. The missing material is known as dark matter.

> Astronomers looking at how galaxies spin around realized that dark matter must exist. The galaxies move faster than they should, and that means they are heavier than the mass of the visible stars. The rest of the galaxy is dark and invisible.

Mystery Matter

This chart, below, shows that there is around five times more dark matter (in purple) than ordinary matter (in green). But there is also something even more mysterious called dark energy. Dark energy was discovered in 1998, and makes up around two-thirds of the energy in the Universe. It is making space expand, but physicists do not really understand it yet.

Dark matter 27%
Ordinary matter 5%
Dark energy 68%

Dark matter detectors are deep underground to shield them from all other particles and radiation.

Detectors

Dark matter is dark because it does not produce light or any kind of radiation. Nor is it affected by electromagnetism in any way. The only force that seems to affect dark matter is gravity. The search for dark matter is very difficult and no one has found any yet. Some dark matter detectors are looking for tiny particles that might be all around us now but do not show up in standard tests.

DID YOU KNOW? One suggestion is that dark matter is the influence of gravity from other universes leaking into our own. No light or regular matter moves between universes, but it affects the apparent mass of matter in our universe.

Astronomers think that some dark matter might be dark objects in the halo around the edge of a galaxy.

If a galaxy was made only from the kind of regular matter that we can detect, all of its stars would be flung out into deep space as it spun around, because it wouldn't have enough mass to keep it together. Heavy dark matter inside must be pulling the galaxy together and keeping the stars in place.

HALL OF FAME:
Vera Rubin
1928–2016

In 1979, after studying the stars in Andromeda, the nearest big galaxy to our own, Vera Rubin showed that dark matter makes up a lot of the mass of galaxies. She calculated the speed at which stars were moving as they went around the center of the galaxy. She found that stars at the edge were moving much too fast for the known mass of galaxy. She calculated that galaxies must have between five and ten times as much mass as we can see.

Chapter 5: Energy and Motion
Doing Work

To a physicist, the term "work" has a very particular meaning: Work is the transfer of energy from one object to another. Work can only happen when a force is applied. Work is measured using units of energy called joules.

In its simplest form, work is using a force to move a mass over a distance. Lifting these sacks is hard work!

Heating Up

When work is done, some of the energy being transferred will become heat. That heat energy is no longer usable and leaks away from the system. This leaking of energy is due to a process called entropy, which means energy tends to spread out. As a result, any working system, from a machine to your body, will always lose energy and eventually stop working unless more force is applied.

Exercise makes us hot because our muscles warm up as they work hard.

Falling Down

The water in this waterfall is doing work as it gushes downward. At the top, the water has potential energy. This is energy it has as a result of its position or state, which can be converted to another form. When gravity causes the water to fall over the cliff, the water's potential energy is converted into kinetic (motion) energy.

The water is a little bit hotter at the bottom of the waterfall because some of the kinetic energy is converted into heat energy.

96

HALL OF FAME: James Joule 1818–1889

The unit of energy, the joule, is named after this British scientist. He wanted to show that heat and motion are related kinds of energy. In the 1860s James Joule performed a famous experiment where he used a falling weight to spin a stirrer in a tank of water. He showed that dropping the weight over and over again made the water gradually get warmer.

The unit of energy, the joule (J), is calculated by multiplying force by distance. So 1 J is the work done when 1 kg (2.2 lb) is moved 1 m (3.2 ft) in 1 second.

The energy from the muscles is transferred to the sack, so it moves up from the ground to the back of the truck.

DID YOU KNOW? The average human body uses about 10 million joules in 24 hours.

Thermal Energy

Physicists call heat energy "thermal energy." It is held by the atoms and molecules in a substance. As the atoms gain thermal energy, they vibrate faster, and the material becomes hotter. Thermal energy is always on the move. It spreads out from hot things to colder things until eventually everything is the same temperature.

The metalworker wears thick, insulated gloves that prevent the heat passing through and burning the skin.

Heat Transfer

There are three ways thermal energy moves—conduction, convection, and radiation. Conduction happens in solids as atoms jostle one another. Convection is caused by warm liquids and gases rising up through colder ones. As hot material cools and falls again, it is replaced by newly heated matter rising, setting up a current that spreads heat. Radiation carries the energy as invisible heat rays, called infrared waves.

Conduction

Convection

Radiation

As the warm water rises, it begins to cool and then sinks. This creates a circular current that moves the heat around.

Temperature scales fix an upper and lower point and divide the gap between into units called degrees. This thermometer shows that the temperature is at the freezing point of water.

Taking the Temperature

Temperature is a measure of the average kinetic energy of the particles in a substance. Temperature is not a measure of the total energy in the material. An iceberg, for example, contains many more atoms than a hot spark coming from a fire, so it carries much more energy in total. However, the spark is at a much higher temperature. Thermometers are used to measure temperature.

Heat is passing along the red-hot iron by conduction. The hot atoms hit their colder neighbors, making them move around more—and become hotter.

An expert can tell the temperature of the metal from the color of its glow.

HALL OF FAME: Lord Kelvin
1824–1907

Born William Thomson, this Irish physicist worked on heat and other forms of energy. He not only transformed the understanding of physics but also helped create new technologies, such as refrigeration and telecommunications. Kelvin discovered that the minimum possible temperature, or absolute zero, is −273°C (−459°F). This is the temperature at which matter has the smallest possible amount of thermal energy. It is now known as 0 Kelvins, with each Kelvin equivalent to 1°C.

DID YOU KNOW? Half the energy from the Sun is invisible infrared heat waves. Light makes up two-fifths of the Sun's energy that reaches the surface of Earth.

Kinetic Energy

The energy of motion is called kinetic energy. Kinetic energy is most obvious when it is making big things like cars and trains move. The Moon, Earth, Sun—and all other objects in the Universe—are moving as well. Even at the smallest scale, moving atoms have kinetic energy.

> The winning cyclist will be the one who is best at converting their muscle energy into kinetic energy. The legs push up and down on the pedals, a motion that is converted by the chain to a rotational motion of the wheels— and so the bike rolls forward.

Speed and Safety

As an object's velocity increases, so does its kinetic energy—but not by the same amount. When speed doubles, the kinetic energy increases by four times. So to go at twice the speed an object needs four times as much energy. This is why speed limits on roads are important. Even small increases in speed mean that vehicles are moving with much more energy and can do more damage in a collision.

The speed limits for heavy vehicles like trucks are lower than for cars because trucks carry more kinetic energy and would be more dangerous in an accident.

Energy Transfer

When one moving object collides with another, the faster object transfers kinetic energy to the slower one. As well as being linked to velocity, the kinetic energy is proportional to the mass of the moving body. So if two objects are moving at the same speed, the larger one always has more energy. How much energy is transferred in collisions depends on the angle of the collision and how hard or flexible the objects are.

The tackler absorbs the kinetic energy of his opponent, stopping him.

The wheels on a road-racing bike are able to grip the road so no energy is lost to slipping. The wheels get hot as they turn fast.

Bicycles are the most efficient means of transport. They convert more of the energy applied into kinetic energy than other vehicles.

HALL OF FAME: Émilie du Châtelet 1706–1749

This French scientist and mathematician linked the idea of kinetic energy to mass and velocity. She also helped to explain that energy of any kind is never made or destroyed, but transformed into other types. Much of Émilie du Châtelet's work is found in her translation of Isaac Newton's writings, to which she added her own ideas and improvements.

DID YOU KNOW? The word "kinetic" comes from the Greek word for movement. The word "cinema" has the same origin—it is a place with moving pictures.

101

Potential Energy

While kinetic energy is concerned with motion of all kinds, potential energy is how energy can be stored by objects in different ways. This energy is transferred to objects when work (see page 96) is done to them. The energy is always there, even when the object seems to be doing nothing—and it can be released when the conditions are right.

Electrical Potential

One form of potential energy is electrical potential energy. This is the energy stored in batteries. It is created whenever a difference in electrical charge is created, where positive charges and negative charges are kept separated. To rebalance that charge and release the electrical potential energy, electrons move as an electric current (a form of kinetic energy).

To recharge a battery, an electric current is used to push charged particles apart to create a store of potential energy. When the battery is used, that potential is released.

> The riders on the rollercoaster feel the sudden release of gravitational potential energy. After being lifted slowly to the top of a hill, they then roll down the other side under the pull of gravity.

Elastic Potential

Some solid materials will deform (change shape) when a force is applied to them. Permanent changes are called plastic deformations. Energy is not stored in the new shape. Temporary changes are called elastic deformations, and the shape will spring back to normal when the force is released. The energy stored in the stretched material is elastic potential energy. This kind of energy makes balls bounce and is used in spring-loaded and wind-up devices.

> Adding elastic potential energy to this bungee helps to work the muscles and keep them strong.

DID YOU KNOW? Rivers flow due to the water's gravitational potential energy. Rain falls high on mountains and the water is pulled back to the ocean by gravity.

HALL OF FAME: Sophie Germain
1776–1831

As a woman growing up at the end of the eighteenth century, Sophie Germain was not allowed to study at a university. Instead, she wrote to many of the world's best scientists and helped them with their projects. In 1816, she won a grand prize given by the Paris Academy of Sciences to understand how elastic solids worked. She was the first woman to be honored by the Paris Academy.

It takes a lot of work to push the rollercoaster up to this point. That work is converted into gravitational potential energy, which is released as the cars roll down the other side.

Gravity is making the cars accelerate at a similar speed to if they were in free fall through the air. The curved track will then slow the cars down gradually.

Other Types of Energy

One of the main laws of physics is the conservation of energy. This law says that energy is never created or destroyed. Instead, it is transformed from one type of energy into another. As well as motion, heat, and potential energy, there are several other kinds of energy.

Electrical Energy

Most modern machines are powered using electrical energy. This is a flow of energy that is carried by a current of electrons. A very simple electrical machine is a heater, where the flow of electrons pushes on atoms in a wire, converting their electrical energy to heat (and perhaps visible light, too). Electric motors use magnets to create forces that make moving parts spin.

The microchips inside a computer are powered by electricity. They use the energy to perform calculations and follow instructions. The computer screen is also lit by electricity.

A megaphone has a cone that wobbles back and forth very fast. That wobble creates a sound wave—a loud one!

Sound Energy

Sound is a wave of pressure that moves through the air. It is a kind of kinetic energy because the sound wave is moving the air molecules, causing them to spread apart and then squeeze together. Sound waves are created by kinetic energy from other materials being transferred to the air. Loud sounds contain more energy than quiet ones.

Energy is released when bonds between atoms break. Some of that free energy is then used to rearrange the atoms into new substances.

Chemical reactions involve chemical energy. The energy is stored in the links or bonds between atoms. This is a kind of potential energy.

If any energy is left over it is often released as heat or light.

HALL OF FAME: Laura Bassi
1711–1778

Born in what is now Bologna, Italy, Laura Bassi became only the second woman in the world to earn a PhD, or doctorate. She was also the first woman to have a paid job as a teacher at Bologna's university, where she taught physics. Bassi studied electricity and became the main researcher in the physics department at the university.

DID YOU KNOW? Thinking uses energy. Your brain uses up 20 percent of the energy your body needs each day.

Power

Work is a measure of the transfer of energy from one object to another. Power is a measure of how quickly the energy is transferred. For example, it takes the same amount of work for two swimmers of the same mass to swim a length of a pool, but a more powerful swimmer can do that work in less time.

This immensely powerful truck can move a very large load in a short amount of time.

Watts

Power is measured in units called watts (W), named after James Watt. He used the idea of power to explain why his steam-powered inventions were able to work better than human workers. One watt (1 W) is 1 joule (1 J) of energy transferred in a second. Most of our appliances at home have a power rating expressed in watts. This tells us how much energy they will use up. A machine with a lower power rating saves energy but might take longer to do the job.

The brightness of light bulbs is measured in watts. Floodlights are around 10,000 times more powerful than the lights in your home.

HALL OF FAME: James Watt
1736–1809

This Scottish engineer is known for his work on the steam engine. He improved on earlier designs to create machines that were efficient to use. Before James Watt, steam engines used a lot of fuel but were not very powerful. Watt's engines were large and powerful and were meant to be used in factories and mines. Later inventors made smaller engines for ships and trains.

This truck is a complex machine that uses several kinds of simpler machines, such as wheels, levers, and pulleys, to do work efficiently.

Using Power

A machine can be more powerful than a human body. This means one machine can do the work of many people. The original unit of power was horsepower. Early machines were compared to the strength and power of horses, which were used to haul heavy loads. The power of cars and other vehicles is sometimes still measured in horsepower today. One horsepower is about 740 W, and it is estimated that one horse can exert 10 times as much power as a person.

This forklift can lift more weight than a person and it can do it faster and go higher, too.

The dump truck uses powerful pistons to push up the load so that it slides off the back.

DID YOU KNOW? The most powerful rocket engine ever built, the F-1 that launched NASA's Saturn V Moon rockets, had a power rating of 41 million watts.

Levers

There are six types of simple machines that are often combined in mechanisms. A machine is any system that carries out work, converting an effort (a force applied by user) into a load (a force applied by machine, e.g. a lever). To be useful, a machine reduces the effort needed to do something. This is called providing a mechanical advantage. The simplest machine is a lever.

> The long oar is a lever that pushes the boat along. The fulcrum is the oarlock where the oar connects to the boat.

Three Classes of Lever

A lever is a stiff rod that moves around a fulcrum—the turning or balancing point. A first-class lever, such as a seesaw, has the fulcrum in the middle, with the effort and load forces on either side. Pushing one side down (effort) makes the other side go up (load). A second-class lever, like a wheelbarrow, has the fulcrum at one end and the effort at the other. The load is applied in the middle. Tweezers are third-class levers, where the effort force is applied between the fulcrum and the load at both ends.

This wheelbarrow is a second-class lever. The gardener's effort lifts the handles. The lever turns at the wheel, and the load in the barrow is held in the middle.

Scissors are a pair of first-class levers connected at the fulcrum. Each lever has a cutting edge that slides past the other.

Changing Forces

A lever is a force multiplier, which means the size of the effort and the load forces are different. A lever can't change the work done to be greater than the effort, but it can exert a larger force over a shorter distance. This is what happens when you use a lever to lift something heavy. You could not move the large object directly, but by moving one end of a lever a long distance you can convert the effort into enough energy to move the object a small distance at the other end.

The rowers pull the handle forward and the oar blade at the other end moves backward through the water. The effort force on the handle is translated into a load that pushes the boat along.

The effort force is applied nearer to the fulcrum than the load force, and so moves the lever a smaller distance. The effort needed to move a load is related to the distance of the force from the fulcrum.

HALL OF FAME: Archimedes
287–212 BCE

This great ancient Greek mathematician lived in what is now Sicily. As well as making mathematical discoveries, Archimedes also made several inventions, mostly to defend his city from attacks by the Romans. One of these inventions was a lever system that could hook onto a ship and unbalance it so that it sank.

DID YOU KNOW? The oldest known lever-based machine was a balance scale used for weighing objects. It was made in Mesopotamia (now Iraq) around 7,000 years ago.

Ramps and Screws

Other simple machines are ramps, wedges, and screws, all of which share the same feature of have a thin end and getting wider along their length. A ramp (or inclined plane) makes it easier to lift heavy objects, a wedge is a cutting machine, and a screw is a mixture of both.

This screw-shaped machine is called an auger. It is twisted into the ground to dig a hole.

Inclined Plane

Work is calculated as force × distance. Lifting a heavy load straight up in one go requires a large force to be applied for short time. A ramp makes it possible for the same amount of work to be done using a smaller force over a longer distance. A flight of stairs is a kind of ramp. It would be very hard work to move between floors without stairs!

The giant stone slabs of the pyramids of Egypt were probably lifted into place using ramps.

Cutting Edge

A wedge has the same basic shape as an inclined plane, with a thin end and a thick end. The wedge is a force multiplier. When an effort force is applied to the thicker side, that force is concentrated in the thin end, which is often a sharp blade. The concentration of force creates very high pressure, so the wedge can cut through what it touches.

An ax head has a wedge shape that slices through most things.

DID YOU KNOW? The first ever machine was the Stone-Age hand ax, a wedge-shaped cutter that was invented by our prehuman ancestors more than a million years ago!

HALL OF FAME: Anna Maria van Schurman 1607–1678

Anna Maria Van Schurman is best remembered as a poet and artist but she was also the first woman to go to university in Europe. She was highly intelligent and was able to speak at least 14 languages! This allowed Van Schurman to exchange letters with many important people all over Europe. The letters discussed all kinds of things, including the latest ideas in physics. Van Schurman helped many people, women and men, learn new things about the world.

The screw works like a wedge or ramp wrapped around in a spiral. As the screw turns, it cuts into the ground and lifts the loose soil up and out of the hole.

Another name for an auger is an Archimedes screw. As well as digging holes, it can be used to lift water and powders.

Wheels and Pulleys

The wheel is a very famous machine. It was probably first used about 5,500 years ago for making pottery. Only later was it repurposed to move carts and wagons. A pulley is a hoist system that uses several wheels together to make it easier to lift heavy loads.

Mechanical Advantage

The simplest types of pulleys use a single wheel with a rope looped around it. This machine just reverses the direction of a force. If you pull down on one side of the rope—the effort force—the other side pulls up with a load force of the same size. A compound pulley with more than one wheel acts as a force multiplier. A pulley with two wheels can move a load four times greater than the effort. However, the compound pulley must be pulled four times as far for it to lift a load to the same height as the simple pulley.

This movable pulley is used to lift and raise loads attached to the hook.

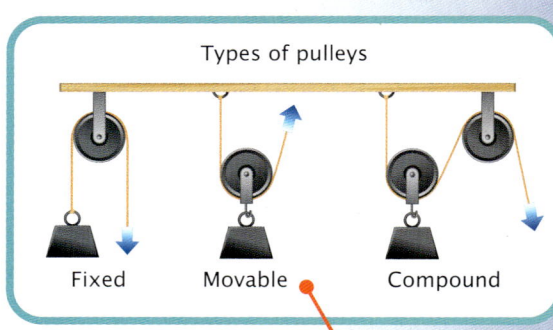

Types of pulleys

Fixed — Movable — Compound

A pulley shares the weight of the load between both sides of the rope, and so less effort is needed to lift it.

HALL OF FAME: al-Jazari 1136–1206

This Arab engineer and inventor is best remembered for building life-like automata. An automaton is a simple kind of robot that can only perform a limited set of movements. Al-Jazari used pulleys and other machines powered by waterwheels to make automata that poured drinks or played music.

Axle Needed

Vehicles need wheels to move. A wheel rotates around an axle, which is a stiff rod attached to the middle of the wheel and connected to the vehicle. In most cases, the wheel turns because rotational motion is transferred to it via the axle. For each rotation of the axle, the wheel turns once. The vehicle moves a distance along the road equal to the distance around the rim of the wheel, which is much longer than the distance around the axle.

The spokes of the wheel spread out from the axle to keep the rim rigid, but make the wheel lighter than if it were a solid disk.

The compound pulley here is attached to cables that can raise and lower the crane's long boom, or arm.

This large pedestal crane is used to lift loads in and out of a ship's hold.

DID YOU KNOW? The largest wheel was a pleasure ride in Dubai. At 250 m (820 ft) across, it was more than twice the length of a football field.

Engines

An engine is a complex machine that harnesses the heat energy released by a burning fuel. It can be used to drive a vehicle over land or water, or through the air or space. Most engines burn fossil fuels and that creates dangerous pollution. In future, many engines will be replaced by electric motors.

A jet engine burns a hot mixture of air and fuel. The gases made by the burning produce a jet that blasts out the back, creating a thrust force.

Heat Engines

The simplest engines are external combustion engines. The fuel is burned outside the engine, and its energy is transferred to it by a working fluid, which pushes on the moving parts of the engine to create motion. The best example is a steam engine. The working fluid is water that is boiled into steam. In an internal combustion engine, like those in cars, ships, and planes, the fuel itself is the working fluid.

Steam engines are very inefficient and dirty. They have not been used for many years.

Gears

The motion of the engine is transmitted to the wheels of a car or a ship's propeller via a series of gears. The gears are interlocking wheels. When one turns, the gears touching it will also move, only they turn in the opposite direction. The number of turns the gears make depends on the number of interlocking teeth. If an adjoining gear has half as many teeth as the main wheel, it will turn at twice the speed.

Gears are used to manage the forces coming from the engine.

HALL OF FAME: Nicolas Léonard Sadi Carnot
1796-1832

This French soldier and engineer investigated how steam engines work in the 1820s. What he discovered could be used to explain how all engines work and how energy is transferred. Nicolas Léonard Sadi Carnot is sometimes called the father of thermodynamics, a field of physics that deals with heat, work, and energy.

The jet engine works by converting chemical energy into the kinetic energy of fast-moving gases.

The flow of gas also drives fan-like turbines that draw in more air and squeeze it to boost the burning process.

DID YOU KNOW? The first engine was the aeolipile invented by Hero of Alexandria 2,000 years ago. It was a simple steam engine that spun around powered by jets of hot steam.

First Law of Motion

The way objects move—how they stop, start, and change direction—is governed by three simple laws. The first law of motion is that an object will resist changing its state of motion until a force is applied to change it.

In snooker, as the cue ball collides with a red ball, it applies a force. The force alters the motion of the balls on the table. The cue ball slows down and changes direction, while the red ball starts to move.

Adding Force

A state of motion can include being still and not going anywhere—what physicists called being at rest—or moving in a straight line at an unchanging speed. These states will carry on forever unless a force is applied. On Earth, the forces of gravity, friction, and drag will always act on a moving object to change its state of motion—and make it stop.

This car jumps as it maintains its motion and continues in a straight line—only there is no ground underneath it. The force of gravity will pull it back down.

HALL OF FAME: Mo Di
Fifth Century BCE

Almost nothing is known about the figure referred to as Mo Di or Mozi, the author of 71 chapters of an ancient book in Chinese. He is thought to have come from a humble family and traveled around trying to persuade warlords to stop fighting one another. Much of his philosophical book has been lost, but the surviving text includes the first known statement that something will keep moving unless a force stops it—the first law of motion that Isaac Newton described in Europe 2,000 years later.

A car's body is designed to absorb the energy of a crash, so less force reaches the passenger compartment. The airbags reduce the rate at which the passengers slow to a stop.

Inertia

The property of matter that makes it resist changes to its motion is called inertia. Inertia means it is hard to get an object moving, and once it is moving it is hard to get it to stop. In a car crash or a fall from a height, forces create a sudden change in motion—a sudden stop against a stationary object. The inertia of the moving body resists that change, and that causes injuries in a crash.

The soft and smooth green surface creates little friction, so the balls roll long distances before they stop.

Some of the force applied from the cue ball to the first red ball that the cue ball hit also applies to any other balls that red ball goes on to hit. The balls that received no force from the impact stay exactly where they are.

DID YOU KNOW? Force is measured in units called newtons. A force of 1 newton is enough to accelerate a mass of 1 kg (2.2 lb) by 1 m (3.3 ft) per second every second.

Second Law of Motion

> A human cannonball uses the second law of motion to make sure the force applied by the cannon will be just enough to throw their body to the safety mat.

The second law of motion concerns the relationship between the mass of an object, the size of the force that acts on it, and the acceleration that the force creates. The force used to move the body is calculated by multiplying the body's mass by its acceleration.

Mass and Motion

All three features of a moving body—the mass, force, and acceleration—are proportional to one another. By increasing the force, a body will accelerate faster. When the mass of the body is increased, a larger force is needed to achieve the same acceleration. In the real world this law explains why it is easy for a person to push a bicycle along the road, but cannot easily produce the force needed to push a car in the same way.

A rescue tow truck is powerful enough to generate the force needed to move itself and the broken-down car.

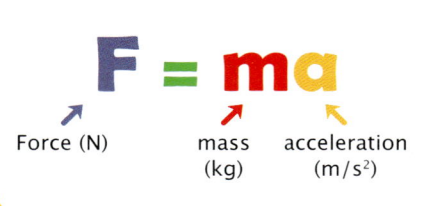

$$F = ma$$

Force (N)   mass (kg)   acceleration (m/s²)

This equation is one of the simplest and most useful in physics.

Calculating Force

The second law of motion can be defined by the simple equation: Force = mass × acceleration. This equation can be rearranged so any of the three variables can be calculated from the other two. For example, the acceleration of a body is calculated by dividing the force by the mass. The mass of the body can be calculated by dividing the force by the acceleration.

The cannon accelerates the human cannonball using a spring-loaded piston—not gunpowder!

The human cannonball is accelerated fast enough to override the pull of gravity, so the person flies through the air.

HALL OF FAME:
Galileo Galilei
1564–1642

This Italian scientist was one of the most important physicists of all time. He made many discoveries about the Moon and planets using his own improvements on the telescope, and carried out experiments on how objects move, especially how they fall. Galileo correctly predicted what would happen if a heavy and a lighter ball were dropped from the Leaning Tower of Pisa—that they would hit the ground at the same time.

DID YOU KNOW? On Earth the acceleration due to gravity is 9.8 m/s^2 (32 ft/s^2). That means every second that gravity is applied, the object's speed increases by 9.8 m/s (32 ft/s).

Third Law of Motion

The third law of motion is perhaps the most famous. It says that for every action there is an equal and opposite reaction. The term "action" means that a force is being applied. The "reaction" created is a force that is equal in size but pushes in the opposite direction.

> It is the third law of motion that makes bumper cars so much fun. Driving into other cars gives everyone a bump!

Rocket Engines

A rocket works by using the third law of motion. The propellants are mixed and burn with great ferocity, and the flames and gas they produce are pushed out of the bottom of the rocket. The action force caused by the rocket pushing away the gases creates a reaction force, where the gases push away the rocket. As a result, the spacecraft lifts off the ground and is powered into space.

The downward force of the exhaust gases made by the burning fuel creates an upward thrust force.

HALL OF FAME:
René Descartes
1596–1650

This great French philosopher is most famous for saying, "I think, therefore, I am." This was his explanation of how he knew he existed as a real person. René Descartes also invented graphs and coordinates, among many other things. He came up with three laws of motion a few decades before Isaac Newton, though they were not as easy to understand as Newton's later versions.

DID YOU KNOW? NASA used the laws of motion to knock an asteroid off course. They simply smashed a spacecraft into it! This system will be used to make sure big asteroids do not hit Earth in the future.

Action and Reaction

The third law explains why pushing on an object makes it move. The force of a skateboarder's foot pushing on the ground creates an opposite force of the ground pushing on the foot (and the skateboard) so the skater moves forward. To slow down, the rider pushes the skateboard tail into the ground. The ground pushes back—and the skateboard stops.

Skateboarders are using the laws of motion.

The rubber bumper stops the forces of the collisions from damaging the cars.

The mass of each car is about the same and their top speed is low, so they hit each other with a small and more or less equal force.

Momentum

Momentum is a measure of how much motion an object has. It is calculated by multiplying the mass of an object by its velocity (speed in one direction). So a heavy truck will have more momentum than a lightweight car traveling at the same speed. The truck and the car will both have more momentum if they travel at a higher speed, and less momentum if they travel at a lower speed.

Angular Momentum

The point around which a skater spins is called the axis of rotation.

This spinning skater makes use of momentum in a circle, called angular momentum. While she spins on the spot, she can speed up and slow down the spin in a simple way. Spreading her arms spreads out her mass, which makes her spin more slowly. Drawing in here arms pulls her mass to the middle, which makes her spin faster. Outstretched arms increase the inertia in the spin, so it slows down. Drawing the arms in decreases the inertia. The momentum stays the same, so the spin speed increases.

Momentum Conserved

Until a new force is applied to a moving body (or a group of objects), its momentum will always stay the same. A helical spring toy is a good example of this. A small push will set the flexible spring tumbling down the stairs. At each step, the spring tips over and begins the tumbling process again because it still has momentum. It only stops at the foot of the stairs, where the force of the flat ground brings it to a halt.

This helical spring toy has very little friction interfering with its motion.

HALL OF FAME: Jean Buridan 1301–c.1360

This French scientist was one of the first people to come up with the idea of momentum. He called it impetus, and recognized that something could continue to move even after a force was no longer being applied to it. Jean Buridan said that impetus could be calculated as weight × velocity. He thought that impetus was what caused motion, but now we understand that forces do that.

This set of swinging balls is called a Newton's cradle. It is a toy that shows how the momentum of a moving system is always conserved.

The moving ball will hit the next ball and transfer its momentum to it. That momentum moves through the middle balls to the farthest one, which will swing up and out.

The middle balls stay completely still while those at the ends swing back and forth. This is because all of the momentum is transferred through the middle balls to those on the ends.

DID YOU KNOW? Supertankers, the world biggest ships, have such a lot of momentum that it can take them 20 minutes to slow to a stop!

Velocity and Acceleration

When measuring motion, physicists use three basic measurements—time, distance, and direction. Together, these can be used to calculate velocity, which is the distance traveled over time in a certain direction. When forces are applied, they create acceleration. Acceleration is a change in the velocity, increasing or decreasing it.

A dragster's enormous engine is designed to push the car forward with as much force as possible. The car will be acelerating throughout the whole race.

The Need for Speed

Speed is a simple measure of how fast an object is moving. Velocity is a speed in a particular direction. Objects traveling at the same speed but in different directions have different velocities. This is especially important when they are traveling toward each other. If their speed is the same, the relative velocity (how fast one is moving in relation to the other) will be twice the individual object's speed.

This runner's speed is calculated by measuring the time it takes for him to move a known distance.

HALL OF FAME: Katherine Johnson 1918–2020

Katherine Johnson was a mathematician working at NASA when the first rockets were being launched into space. It was her job to calculate the flightpaths that rockets took as they flew up into orbit. John Glenn, the first American astronaut to orbit Earth, asked Johnson before his flight to check the velocity calculations made by the computer. When she said they were correct, Glenn was happy to launch.

DID YOU KNOW? The fastest speed in the universe is the speed of light through a vacuum. That is 300,000 km per second (186,000 mi per second).

A drag race runs along a straight course. The winning car is the one that can accelerate fastest.

Making the car so long helps it stay on the ground as it zooms along.

Steering around a bend is a form of acceleration.

Changing Speed

Acceleration occurs when a force is applied to move an object. It is a measure of the rate of change of velocity. Acceleration is expressed as how much the speed will increase or decrease every second that the force is being applied. When the force stops, the acceleration also stops and the velocity remains constant. A force may keep the speed unchanged but change the direction of the object—and thus change its velocity. This is also acceleration.

Chapter 6: Waves and Optics

Properties of a Wave

> Waves have high peaks and low troughs. The wavelength is the distance from one peak to the next.

Many forms of energy travel as waves. A wave is a vibration that transfers energy from one place to another without transferring matter. All waves have wavelength, frequency, and amplitude. Although light and sound are very different, because they are both waves they behave in predictable ways. The properties of a wave determine how it appears to us—as a high sound or a red light, for example.

Frequency

The frequency of a wave is how many complete oscillations, or cycles, the wave makes every second. Frequency is measured in hertz (Hz), and a wave with a frequency of 1 Hz completes one oscillation in one second. Frequency is a good indication of the energy in a wave. High-frequency waves carry more energy than low-frequency waves of the same type.

These instruments make sound waves with different frequencies. Our ears detect high-frequency sounds as high-pitched notes, while low-frequency notes sound deeper.

HALL OF FAME: Pythagoras c.570–c.495 BCE

This ancient Greek mathematician is most famous for his work on right-angle triangles. However, among many other discoveries, Pythagoras found a link between the note made by a string on a musical instrument and the string's length. Halving the length of the string moves the note up one octave.

126

Speed

A wave's speed is calculated by multiplying its wavelength by its frequency—so if the wavelength is 0.5 m (1.6 ft) and the frequency is 5 Hz, the speed is 0.5 × 5 = 2.5 m per second (or 1.6 x 5 = 8.2 ft per second). Sound waves and ocean waves can travel at different speeds, but the speed of light is always the same when it travels through the same medium. It is fastest in the vacuum of space, and slows down slightly in air and even more in water.

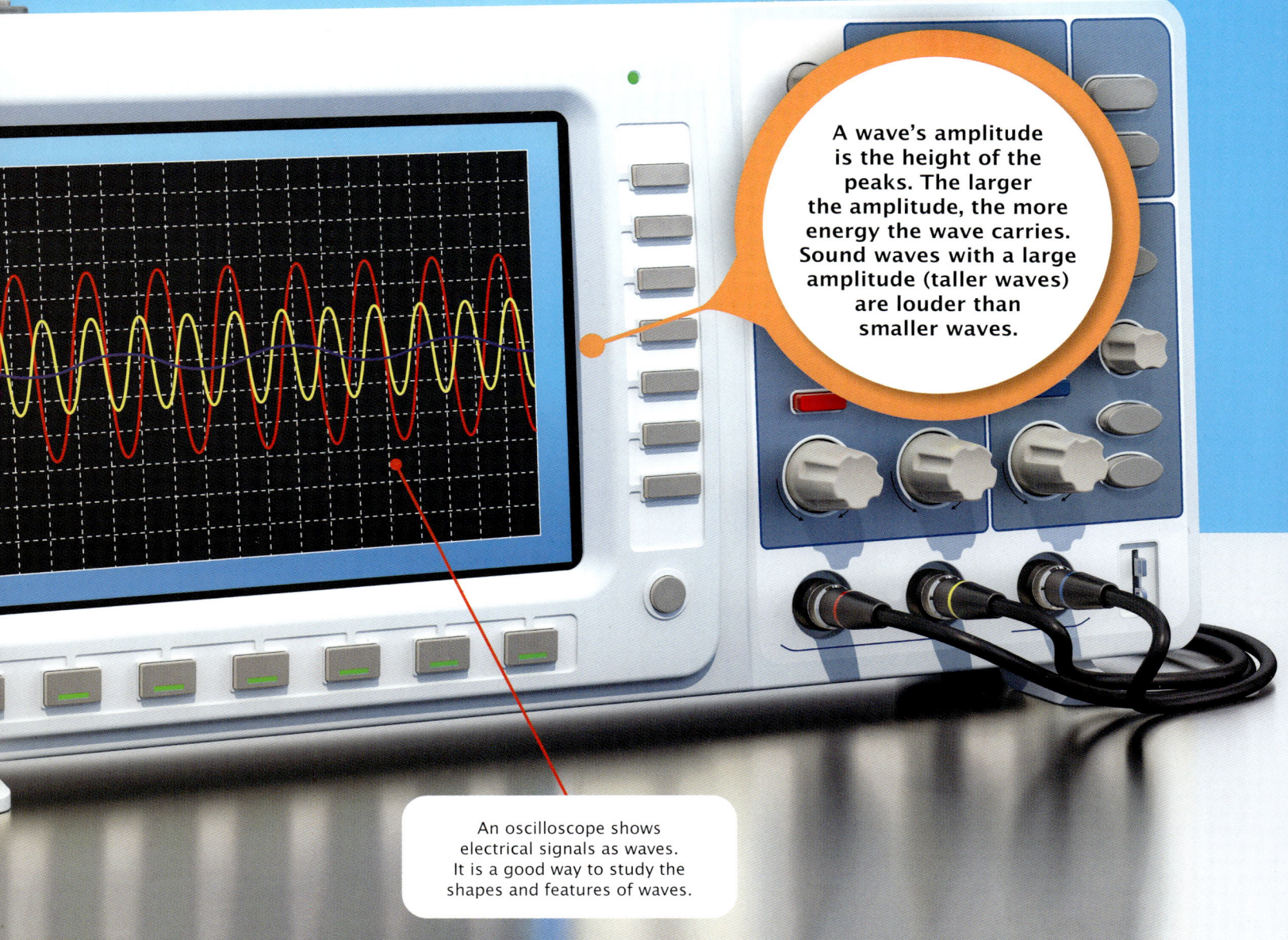

This jet fighter is punching through the sound barrier, creating a puff of cloud as it moves faster through the air than sound waves can.

A wave's amplitude is the height of the peaks. The larger the amplitude, the more energy the wave carries. Sound waves with a large amplitude (taller waves) are louder than smaller waves.

An oscilloscope shows electrical signals as waves. It is a good way to study the shapes and features of waves.

DID YOU KNOW? Earthquakes are caused by powerful seismic waves that ripple through Earth. When they reach the surface, the waves make the ground shake.

Types of Waves

A wave is the form energy takes when traveling. There are three kinds of waves—Longitudinal, transverse, and surface waves. Each one oscillates in a particular way. Sound waves and some seismic waves (the waves that run through Earth's interior) are longitudinal. Light and other radiation are transverse waves, while the swells of an ocean are surface waves.

Longitudinal waves, such as sound, need a medium to travel through, such as air. There is no sound in space because it is empty—there is no medium.

Surface Waves

A surface wave forms where two substances meet. The most familiar are the waves in the sea, where water and air meet at the surface. The water itself does not move with the wave. Instead, the wave's peaks and troughs form as water moves up and down, cycling in the same place.

Ocean waves break in shallow water because the bottom section of the wave slows as it drags along the seabed and the top half topples forward.

A loudspeaker converts an electrical signal coming from the guitar into a vibration that creates a sound wave in the air. That wave matches the sound made by the guitar—only it is much louder!

HALL OF FAME:
Inge Lehmann
1888–1993

This Danish scientist was a pioneer of seismology, the study of waves moving underground. Inge Lehmann discovered Earth's solid inner core in 1936. She investigated powerful waves made by earthquakes on the other side of the world. These waves traveled through the middle of the planet, and the way some were blocked or reflected revealed to Lehmann that Earth's hot metal core has an inner region made from solid metal.

DID YOU KNOW? The largest known water waves are tsunami, which reach the shore as a flood rather than a towering wave. After a major nearby earthquake, a tsunami can be 30 m (100 ft) high.

Transverse waves, such as light, do not always need a medium. They can move through space.

Different Oscillations

The oscillation of a transverse wave is up and down, while a longitudinal wave vibrates backwards and forwards. That means a transverse waves has a rising and falling wave shape. A longitudinal wave is harder to imagine. It has sections that are squeezing together and others that are bouncing apart.

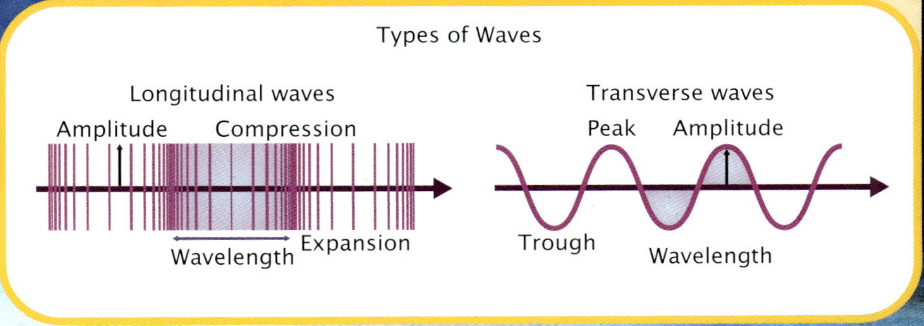

Types of Waves

Longitudinal waves
Amplitude Compression
Wavelength Expansion

Transverse waves
Peak Amplitude
Trough Wavelength

129

Electromagnetic Spectrum

Light waves are just one part of a wider range of waves called the electromagnetic spectrum. Our eyes are able to detect a narrow band of wavelengths in the middle of the spectrum. Electromagnetic waves with different wavelengths are invisible to us—but just as real. As well as light, the electromagnetic spectrum includes radio waves, infrared heat waves, X-rays, and gamma rays.

> Light that appears white is a combination of wavelengths from red to violet light. Bright sunlight looks white.

Ultraviolet

This invisible radiation appears next to violet light in the spectrum. That is why it is called ultraviolet (meaning beyond violet) or UV. UV has a shorter wavelength and carries more energy than the light we see. The UV rays in strong sunlight can burn a person's skin, causing sunburn.

Some paints and dyes glow when UV radiation shines on them.

Radio Waves

Radio waves have very long wavelengths compared to light. They carry only small amounts of energy, which makes them safe to use for communications. Radio and TV signals are both broadcast as radio waves. Microwaves have a shorter wavelength than radio. They are used in telecommunications and in microwave ovens. The energy from a microwave is absorbed by the water in food. That heats the water, which heats the food around it.

> Radio waves carry signals between walkie-talkies as well as between phones.

HALL OF FAME: James Clerk Maxwell 1831–1879

This Scottish physicist was the first person to explain how light and other electromagnetic radiation works. In 1865, he said that radiation is made of waves running through the electrical and magnetic force fields that fill the whole of space. James Clerk Maxwell's achievements in this area are as important as those of Isaac Newton, who explained the light spectrum 200 years before.

Our eyes see light of different wavelengths as different colors. Red light has the longest wavelength; violet light has the shortest. Green and yellow are in the middle of the range we can see.

Sunlight also contains invisible heat waves. These have a longer wavelength than red light and come right after red light in the spectrum. The scientific name for heat waves is infrared, which means "below red."

DID YOU KNOW? Gamma rays have the highest energy of any electromagnetic radiation, with wavelengths less than a trillionth of a meter.

Interference

When two waves meet, they interact. They might combine, cancel each other out, or change direction. This is called interference. It happens with all kinds of waves, from ocean waves and sound, to the radio signals carrying phone calls and TV shows. Interference with these radio waves can change the signal and spoil the sound and picture.

Oil spilled on wet ground creates a rainbow pattern as the light waves interfere with one another.

Adding Waves

What happens when waves meet depends on their phase. When waves are in phase, they are oscillating in sync, rising and falling at the same time. When two in-phase waves meet, they combine to make one bigger wave. This is constructive interference. When waves that are completely out of phase meet, they cancel each other out, disappearing in destructive interference.

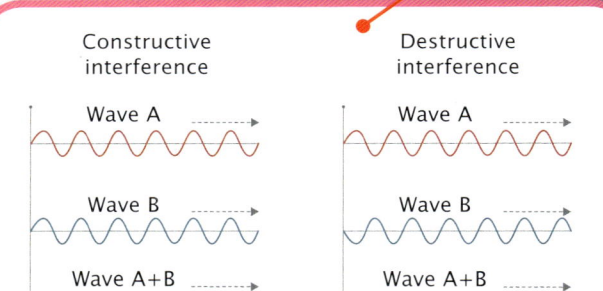

Wave interference

HALL OF FAME: Guglielmo Marconi
1874–1937

This Italian inventor created the first radio communication systems. In 1901, Guglielmo Marconi's radio transmitters and receivers were powerful enough to send a signal all the way across the Atlantic Ocean. At first, messages were sent as Morse code, but later advances in microphones made it possible to send voices using radio waves.

DID YOU KNOW? A Marconi radio was used to call for help from the ship RMS *Titanic* as it was sinking in 1912. Unfortunately, the nearest ship had turned its radio off for the night.

A very thin layer of oil floats on the water. Some light reflects off the oil and other waves pass through it and reflect off the water. The two sets of reflected light are out of phase, so they interfere.

These headphones use interference to cut out unwanted loud sounds.

Some colors are removed by interference and others remain, making this pattern.

Noise-Cancelling Technology

Headphones have a speaker over each ear. However, loud noises from elsewhere can make it hard to hear the music coming through the headphones. To fix this problem, noise-cancelling headphones record these extra sounds, then shift the wave of the unwanted sound by half a wavelength and play it back again. This playback is out of phase with the original sound and the two waves cancel each other out by destructive inference. The unwanted sound wave is gone!

Reflection

Reflections are most familiar from mirrors and other shiny surfaces. A wave hits a surface that it cannot travel through, and so is pushed back the way it came. All kinds of waves can be reflected. An echo is the reflection of a sound wave. We see objects around us only because of the light reflected off them.

The light beams reflected from the smooth mirror are arranged in the same way as the light hitting it. That is why we can see a reflected image.

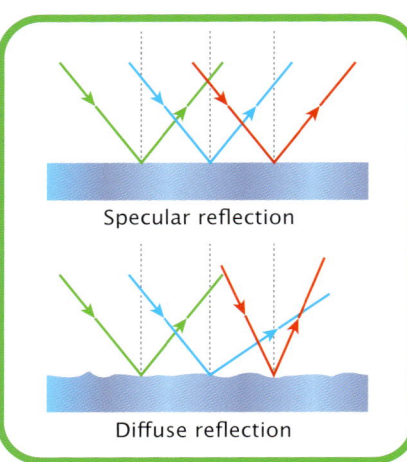

Angle of Reflection

Beams of light always travel in a straight line, and they reflect off a surface at the same angle they hit it. If the surface is smooth, like a mirror, the many light beams that reflect off it all stay aligned—this is called specular reflection. If the surface is rough, as in a puddle when it is still raining, the beams strike it at different angles and so the reflected beams are not aligned, which is called diffuse reflection.

Radar

Radar systems bounce radio waves off objects and monitor the reflections that come back. They can be used to track planes and ships that are too far away to see. Weather satellites use radar tuned to reflect off clouds, especially those that are full of rain.

The radar image tells meteorologists about storms brewing.

DID YOU KNOW? We can only see the Moon and the planets because they reflect sunlight.

134

This funfair mirror has gentle curves, which make the reflection look a bit strange!

The shirt looks green because it only reflects green light. The beams of red, blue, and yellow light are absorbed and not reflected.

HALL OF FAME:
Hasan Ibn al-Haytham
c.965–1040

This Persian scientist was one of the first people to carry out scientific experiments in physics. He made several discoveries in optics, which is the science of light and lenses. Hasan Ibn al-Haytham showed that our eyes detect the light coming from objects. Previously, most people believed the eyes sent out invisible beams that scanned the surroundings and gathered images.

Refraction

Light travels through see-through materials, such as air, water, glass, and clear plastics. When a light beam passes between substances with different densities, it changes direction slightly. This is called refraction. Refraction makes objects appear to bend or become disjointed. It is caused by the light changing speed very slightly as it moves from one substance to the other.

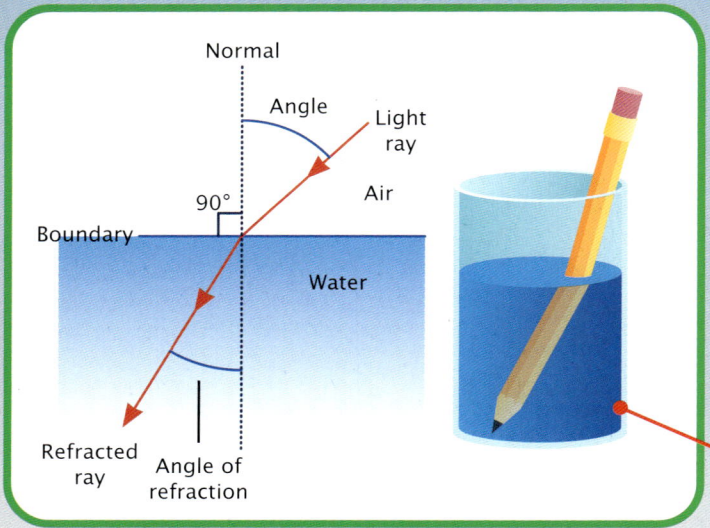

Angles of Refraction

As light moves from air into water, it slows down slightly because water is denser than air. In this beaker, light hits the water at an angle. Since water is denser than air, the light bends toward the normal (an imaginary line at right angles to the boundary between the two mediums) and away from the surface of the water. This makes the lower part of the pencil look bent.

The angles of light beams are always measured from an imaginary vertical line called the normal. The normal is always at right angles to the surface.

Internal Reflection and Refraction

When a beam of light hits the boundary between two substances at a large enough angle, it does not refract. Instead, it is reflected. When sunlight falls on raindrops, the light is refracted as it passes from water to air, then reflected off the back of the raindrop, and refracted farther as it leaves the raindrop. That separates the different wavelengths of light and makes a rainbow.

A rainbow only appears when the Sun is behind you, so the light can fall onto raindrops and be reflected back toward you.

DID YOU KNOW? Starlight is refracted as it shines through different layers of air. This makes the star appear to twinkle in the sky.

HALL OF FAME: Willebrord Snellius
1580–1626

This Dutch scientist set out Snell's Law, which explains how much light will be refracted as it passes from one substance to another. It is sometimes called Ibn Sahl's Law, as it had already been described by Ibn Sahl in Persia (now Iran) in 984. Every transparent material has a refractive index that sets how much it will refract light. The larger the number, the more the light will be refracted.

Refraction makes the pencil appear to shift underwater. The pencil is straight, but the light coming from it is refracted as it passes between air and water.

Light travels more slowly through water than through air.

Because the refracted light is traveling in a slightly different direction than the light coming directly through the air, we see the submerged part of the pencil as shifted slightly to one side.

Lenses

A lens is a curved piece of glass or plastic designed to refract light in a particular way. It can focus light rays into a narrow beam, or spread them out. A camera uses lenses to focus light to capture a sharp image of a scene. Lenses are also used for magnifying small objects or viewing those that are very far away.

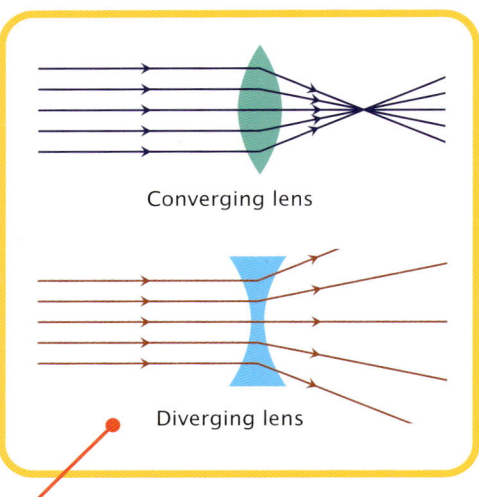

Converging lens

Diverging lens

A diverging lens can be used to correct difficulties seeing far-away things. A converging lens can be used to correct difficulties focusing on near things.

Bending Light

When parallel rays of light hit the curved surface of a lens, they strike it at different angles. As a result, they are refracted by different amounts. A converging lens has a convex surface that bends the rays toward each other. They end up crossing at the "focal point." A diverging lens has a concave surface that causes the rays to spread out.

A single magnifying glass is good for seeing the details of small objects. The best way to use it is to hold the lens near your eye and then move toward the object until it is in focus.

Magnifying

A lens can make an object appear larger so that its fine details are clear. The object is placed close to the lens. The way the rays of light are refracted makes the image of the object appear larger than the original object.

Your eye and brain extend the light rays along the angled path produced by the lens, creating an image that is larger than the actual object.

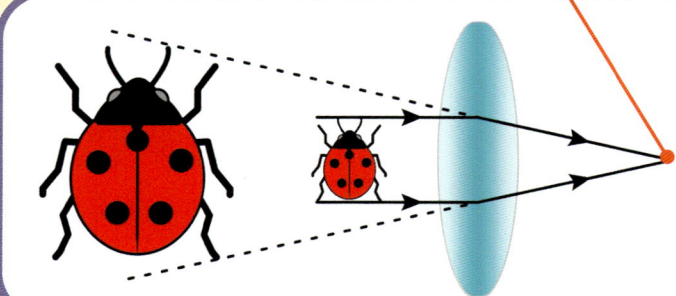

HALL OF FAME: Joseph von Fraunhofer 1787–1826

This German lens maker developed a new kind of glass that gives very clear images. Joseph Von Fraunhofer made the first spectroscope, used for examining the light from flames and stars. A spectroscope uses a lens to split light into a spectrum. By comparing starlight and light from burning substances found on Earth, scientists can work out which chemicals are present in stars.

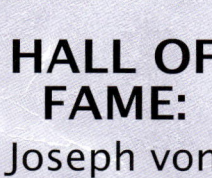

Your eye needs to be at the focal point to see a clear image. This is where the light is focused.

Lenses can focus the invisible heat rays from the Sun. A magnifying glass can be used to start a fire, so be careful when using one.

DID YOU KNOW? Your eye has a lens that can grow fatter or thinner as necessary to focus light on the back of your eyeball. Its size is controlled by muscles in the eye.

Distortion and Diffraction

Over a long distance, waves often encounter obstacles. They may bounce off the object and change direction (diffraction), or they may move around the object. If there is nothing in the way, but the object making the wave is moving, the wave itself can be distorted—stretched or squashed—as it makes its way to your eyes or ears.

Gaps and Obstacles

A wave travels in a straight line until it hits a solid obstacle. If there is a gap in or beside the obstacle, the wave will pass through or around. If that gap is wider than the wave's wavelength, then the wave carries on in a straight line to the other side. If the gap is smaller than the wavelength, then the wave will diffract. It fans out on the other side of the gap, spreading in a circular pattern. If the gap is too small, the wave is reflected instead.

An owl makes use of how sound can move around objects as its call, with a long wavelength, carries through woodland to other owls. The call of songbirds, with a short wavelength, does not spread out.

Squashed and Stretched

If the object making a wave is moving away from you, waves are stretched and the wavelength increases. If the object is moving toward you, the waves are squashed, reducing the wavelength. The light from distant stars moving away from us becomes redder, while the light from stars moving toward us is squashed and becomes bluer. This is called the Doppler effect.

As a swan swims forward, the ripples it makes are pushed close together in front of it but spread out behind it.

Loud emergency sirens can be heard approaching from a long way off. The siren's pitch drops to a lower note when the ambulance drives past and moves away.

The sound wave is squashed as the ambulance comes toward you. Then the wave stretches as the ambulance drives away, making the sound deeper.

The Doppler effect happens with sound, light, and other types of waves.

HALL OF FAME: Rosalind Franklin 1920–1958

This British scientist was an expert in X-ray diffraction. As X-rays pass through tiny gaps between atoms, they are diffracted, making patterns. Rosalind Franklin figured out how the atoms were arranged in large molecules from the diffraction patterns. In the 1950s, she helped work out the shape of DNA molecules. DNA is a complex chemical that carries the instructions for making a living cell.

DID YOU KNOW? Doctors can use the Doppler effect to measure the flow of blood in a living patient.

Chapter 7: Electricity

What Is Electricity?

> This metal dome is charged up with static electricity. The metal is full of charged particles that do not flow in a current, but are static (stay still).

Electricity is a form of energy that results from particles having a charge. When charged particles move in a stream, or current, electrical energy flows and can be used to do work. Electricity is an important source of power for technology, such as computers, heating, and lights. Electricity is widespread in the natural world, too. Electrical charges carried in your nerves make your muscles move, for example.

Moving Charge

Matter can have a positive or negative electrical charge, depending on how many electrons are present. Extra electrons give matter a negative charge. Two objects with the same charge repel each other (push apart), but opposite charges attract each other, pulling together. The attraction between a negative and a positive charge is what pulls electricity along, making a current flow.

This spark of electricity is pulled through the air because one wire has a positive charge and the other is negative.

Electrical Energy

Electrically powered devices use the energy of flowing electricity to do work, transforming it into another kind of energy. A heater or oven converts the electricity into heat energy, while an electric car uses the energy to move. These devices need a constant flow of electricity to keep working, as the energy coming into them is converted into another form as they work.

Food mixers use electricity to spin sharp blades at high speeds that cut food into smaller chunks.

Each strand of hair has the same electric charge, so they repel one another—and create this new hairstyle.

Touching the dome transfers an electric charge to the person. The dome produces very little current, so a dangerous amount of electricity does not flow into the person.

HALL OF FAME:
Nikola Tesla
1856–1943

This inventor and electrical engineer was born in Serbia but moved to America as a young man. Nikola Tesla invented many electrical devices, including an electric motor, a wireless lighting system, remote control that worked by radio, a vertical take-off plane, and a "death ray." He also investigated X-rays and radio waves. He struggled to find money to fund his inventions, and most were never produced commercially.

DID YOU KNOW? The word "electricity" comes from the ancient Greek word for amber. Early scientists found that rubbing amber made it give out sparks.

143

Conductors and Insulators

Electricity requires electrons to move around inside a material. Conductors are materials that carry—or conduct—electricity well. This is because they have plenty of free electrons inside that can move around. Insulators are the opposite, and block the movement of electricity.

These huge cables carry large amounts of electricity over long distances around the country. They are very dangerous if touched and so are strung high above the ground between tall towers called pylons.

Live Wires

Electrical cables are made from a combination of conductors and insulators. The central part is a metal wire, often copper, which is an excellent conductor. The electricity flows through this metal. Around the wire is a flexible plastic coating. Plastic is a very good insulator, and it ensures that none of the electrical energy in the copper leaks out of the wire before it reaches its destination. The insulating layer makes electric cables safe to handle.

The electricity in wires can be powerful enough to kill. Do not touch a cable if you can see the metal inside.

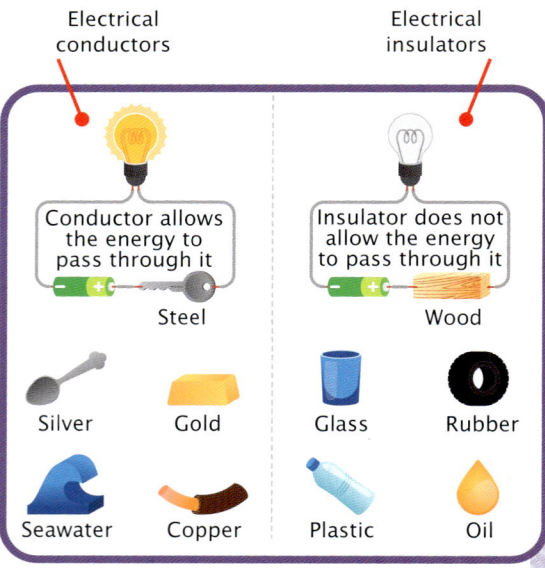

Electrical conductors — Conductor allows the energy to pass through it: Steel, Silver, Gold, Seawater, Copper

Electrical insulators — Insulator does not allow the energy to pass through it: Wood, Glass, Rubber, Plastic, Oil

Carriers and Blockers

All metals can conduct electricity because they have many free electrons inside. Copper, gold, and silver are better conductors than most metals. Seawater is also a conductor because it contains many dissolved salts in the water. These exist as charged particles called ions that can carry electricity. Insulators are nonmetal substances such as plastic and glass. Their electrons are locked in place so they cannot form a current to carry the electrical energy.

144

HALL OF FAME: Stephen Gray
1666–1736

This British teacher had worked as a clothmaker when he was young. Stephen Gray saw that some cloth made electric sparks as it was woven. He experimented with electrical charges, made by rubbing a glass tube with a cloth. He found that the charge from the tube traveled through metals and some threads, but was blocked by ivory and silk. Gray had discovered conductors and insulators.

The wires are kept separate from the metal pylons by insulators like this one made from a glass-like substance.

The workers are using plastic poles to fit the wires in place. The electricity will not flow through this material.

DID YOU KNOW? A superconductor is a material that conducts electricity without losing any energy at all. However, they only work this way when they are very cold.

Current

Electricity can be static or flow as a current. Both occur when there is a difference in electric charge. Static electricity exists when there is an imbalance of electrical charge on a surface. If the surface touches a material that can carry away the extra electrons, a spark appears as the charge is instantly rebalanced. An electric current forms when a difference in charge is maintained, so charged particles flow constantly.

This ball is filled with electrified gas. When an electric current flows through the gas it leaves a trail of glowing plasma.

Moving Electrons

The current moving through metal wires is carried by electrons. Since electrons have a negative charge, they move toward an area with a positive charge. That means the current flows from the negative (−) end of a battery toward the positive (+) end.

When the current is off, the electrons in the wire move in all directions. When the current is on, they all flow in one direction.

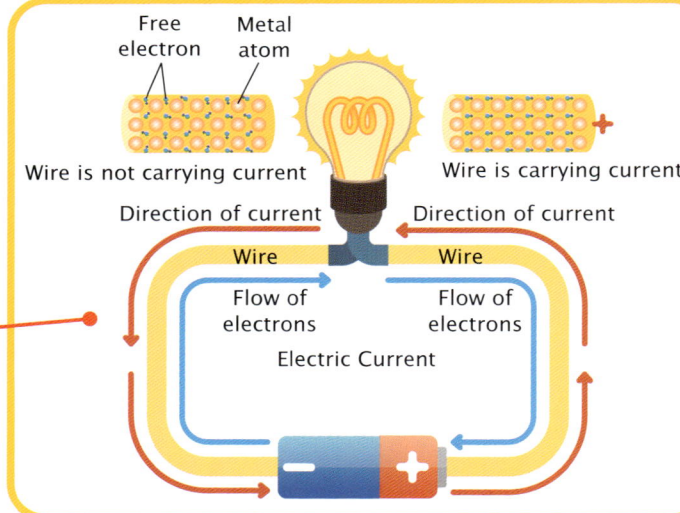

Free electron | Metal atom
Wire is not carrying current | Wire is carrying current
Direction of current | Direction of current
Wire | Wire
Flow of electrons | Flow of electrons
Electric Current

HALL OF FAME: Luigi Galvani
1737–1798

This Italian scientist investigated electric currents—by using frogs' legs! Luigi Galvani was an expert in muscles and nerves. He saw that the muscles of dead frogs' legs twitched when struck by an electrical spark. He thought this was due to a special kind of electricity called "animal electricity," which was later disproved by Volta.

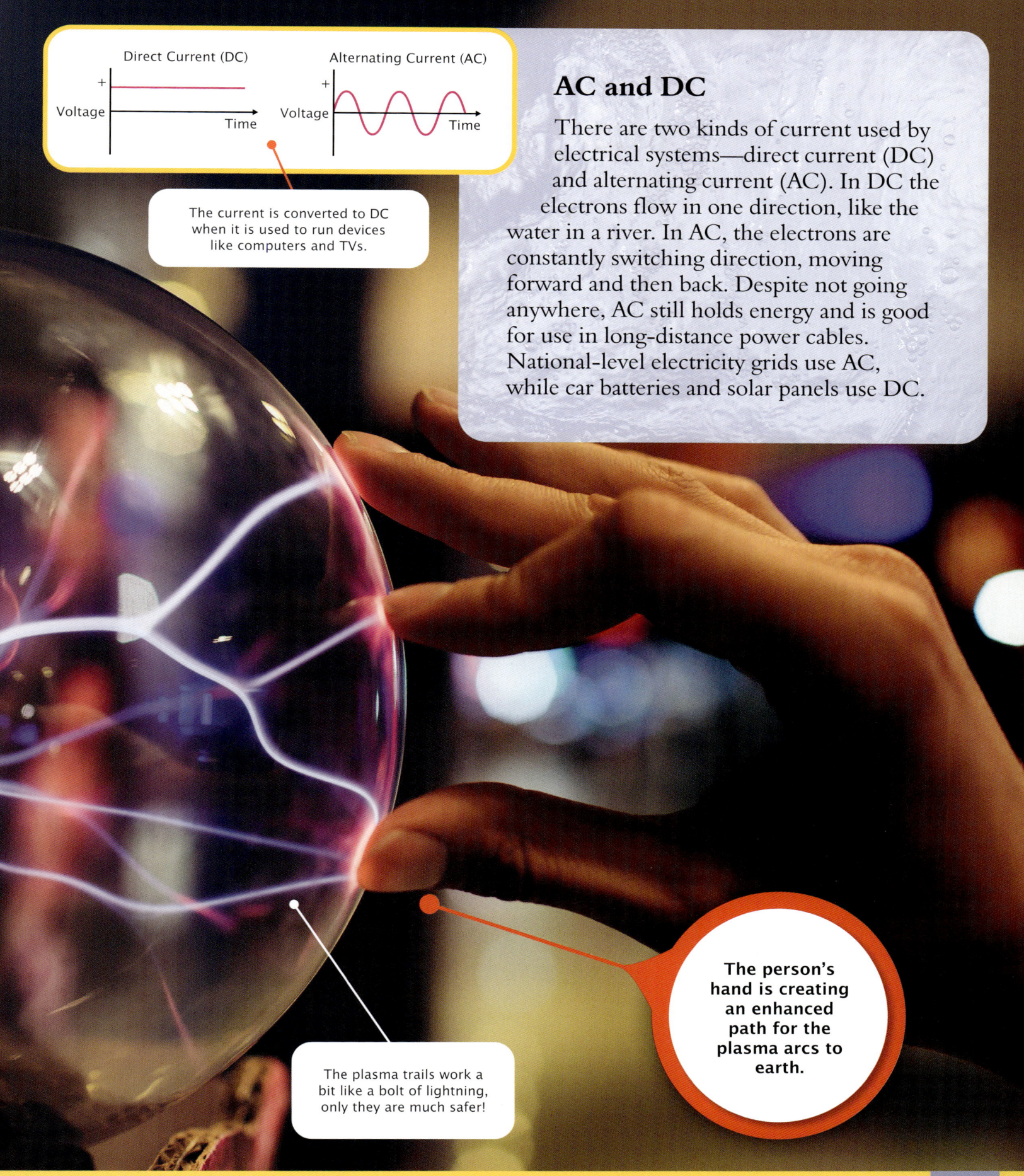

AC and DC

There are two kinds of current used by electrical systems—direct current (DC) and alternating current (AC). In DC the electrons flow in one direction, like the water in a river. In AC, the electrons are constantly switching direction, moving forward and then back. Despite not going anywhere, AC still holds energy and is good for use in long-distance power cables. National-level electricity grids use AC, while car batteries and solar panels use DC.

The current is converted to DC when it is used to run devices like computers and TVs.

The person's hand is creating an enhanced path for the plasma arcs to earth.

The plasma trails work a bit like a bolt of lightning, only they are much safer!

DID YOU KNOW? Electric current is measured in amperes, or amps (A). A current of 1 amp moves 6,240,000,000,000,000,000 electrons every second.

Voltage

Just like anything that moves, an electric current needs a push to getting going. This electrical push is called the voltage. Voltage is a measure of the difference in electrical charge. When the difference is large, the voltage is high and the current moves with great force. A high voltage is needed to get large currents moving.

If the voltage is high enough, it can push an electric current through anything, even air. This is what happens during a lightning strike.

Danger

Electricity is dangerous. Its energy can burn the skin, damage internal organs, and even stop the heart from beating. High-voltage currents can hurt people who just get too close— even if they do not touch—so always take notice of warning signs.

Transformers

The voltage of an electric current is controlled by a transformer. Power plants make low-voltage electricity, which is transformed to a high voltage to be sent over a long distance. Before it enters homes, the electricity is reduced to a much lower, safer voltage.

Transformers are housed in electricity substations, which handle the power supply for a local area.

DID YOU KNOW? The loud thunder crack from lightning comes from the electricity making the air spread out so fast that it breaks the sound barrier.

The lightning is jagged because it is finding the easiest path through the air. Air is normally an insulator.

Lightning bolts are huge sparks that run between storm clouds and the ground. Swirling winds create differences in charge between the sky and the ground, which are then quickly discharged as lightning.

HALL OF FAME: Alessandro Volta
1745–1827

The word "volt" comes from the name of this Italian physicist. In the early 1800s, Alessandro Volta invented the first battery. He used piles of metal disks stacked with acid-soaked paper. These substances reacted with one another, creating a stream of electrons that flowed out of one end and into the other. This is the same system used in today's nonrechargeable batteries.

Circuits

Electric currents only flow through closed loops called circuits. A circuit connects a power supply to electrical appliances and provides them with energy. A circuit can be controlled with a switch to provide or cut the power. When a switch is turned off, the connection between the wires is broken so current cannot flow. When it is on, the circuit is complete and current flows.

Light strings are a simple series circuit. There is one circuit running through all the lights in turn.

In Parallel

Circuits can connect devices in parallel or in series. This is a parallel circuit. Each light has its own connection to the power supply. If one light is disconnected, the other two will stay on as current still flows through the other paths. The voltage is the same in each path of a parallel circuit, but there might be different currents depending on the resistance of each light.

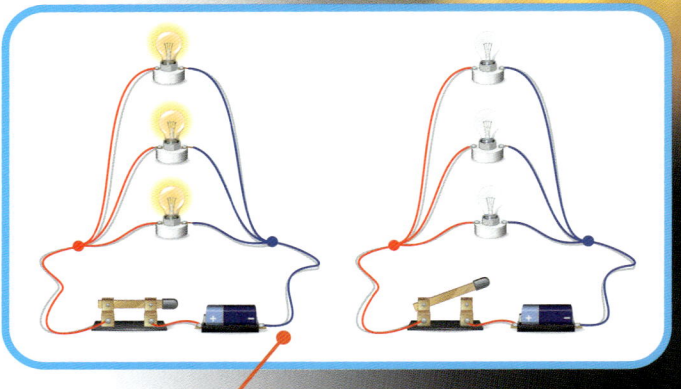

The main switch disconnects all parts of the circuit at once.

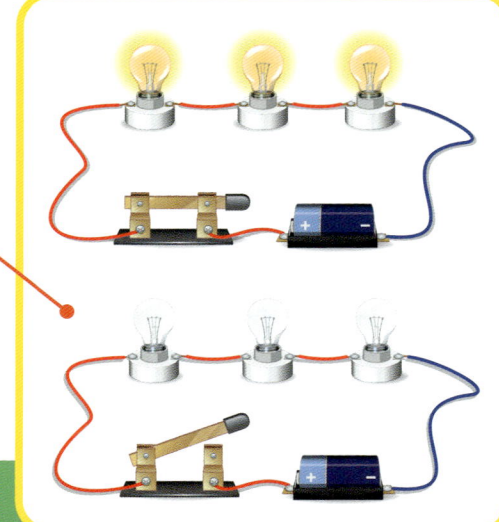

This kind of circuit is easy to create but it is not as useful as a parallel circuit. Most circuits in the home are in parallel.

In Series

These lights are connected in series. They are in a single line on the same connection to the power supply. If the circuit is broken anywhere, all three lights will go out. There is only one path for current and the same current runs through every light.

HALL OF FAME: Edith Clarke 1883–1959

Edith Clarke was the first woman to work as an electrical engineer in the United States. She later became the country's first female professor of electrical engineering. Clarke was an expert in the workings of the power grid and invented a system that made in easier to calculate the voltage, resistance, and other features of the high-powered transmission cables.

This is the simplest lighting system. However, if one light fails they all go out, and it can be difficult to find the faulty one that needs replacing.

The lights are all identical and so they glow with the same brightness because the same current is flowing through all of them.

DID YOU KNOW? The longest single length of electrical cable in the world is 5.4 km (3.4 miles) long. It connects a power station in Greenland to the capital city Nuuk.

Electric Power

The forces of electromagnetism can be used to convert the flow of electricity into motion in an electric motor. The process also works the other way around. Moving magnets and conductors can be used to generate (create) electricity. Electric motors are at the heart of a lot of modern technologies.

The electric current from this charging station was generated elsewhere and brought here by the power grid.

Electric Motor

When electricity flows through a wire, it generates a magnetic field and the wire becomes magnetic. An electric motor uses the attraction and repulsion forces of magnets to make an electric wire spin around very fast. By increasing the size of the current and the strength of the magnets, the motor can be made to spin more quickly, producing enough power to drive a car.

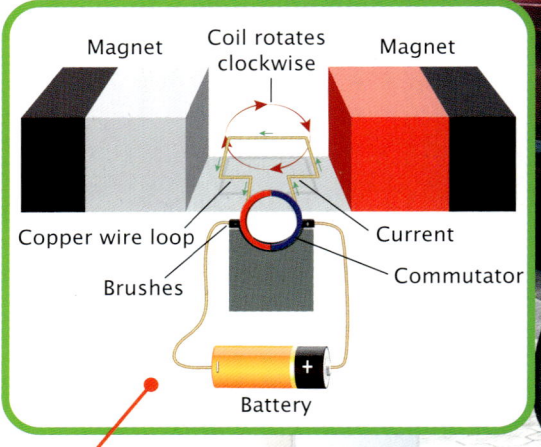

The ring-shaped commutator always directs the current into the right side of the wire so the forces keep pushing it around in the same direction.

HALL OF FAME: Michael Faraday 1791–1867

This British scientist invented both the electric motor and the electricity generator. Although Michael Faraday had little formal education and was self-taught, he became one of the most important scientists of his day. He discovered the relationship between electricity and magnetism, and that light is affected by magnetism. He also worked on electrolysis—how chemicals in liquids can break down when an electric current is passed through them.

152

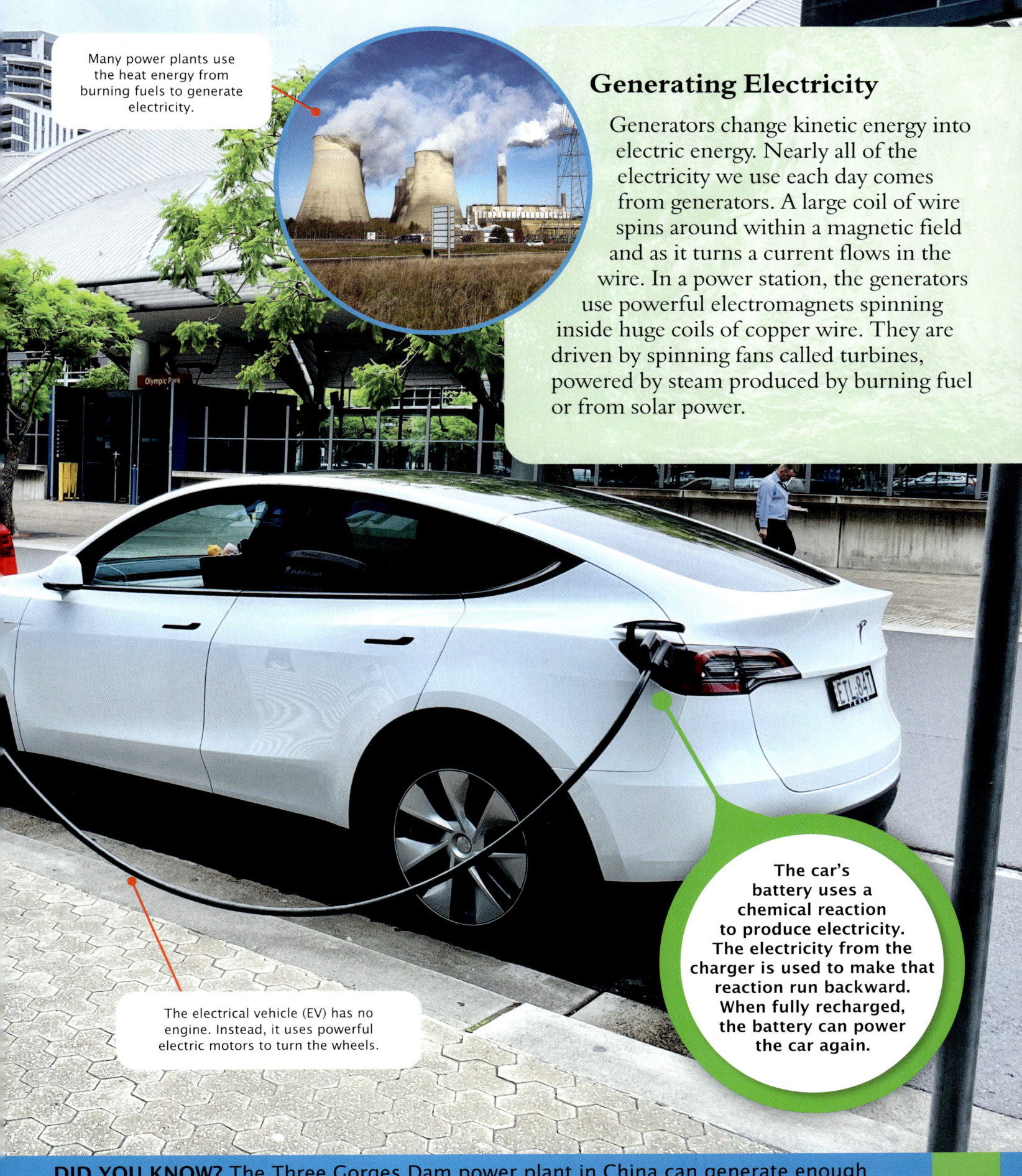

Many power plants use the heat energy from burning fuels to generate electricity.

Generating Electricity

Generators change kinetic energy into electric energy. Nearly all of the electricity we use each day comes from generators. A large coil of wire spins around within a magnetic field and as it turns a current flows in the wire. In a power station, the generators use powerful electromagnets spinning inside huge coils of copper wire. They are driven by spinning fans called turbines, powered by steam produced by burning fuel or from solar power.

The car's battery uses a chemical reaction to produce electricity. The electricity from the charger is used to make that reaction run backward. When fully recharged, the battery can power the car again.

The electrical vehicle (EV) has no engine. Instead, it uses powerful electric motors to turn the wheels.

DID YOU KNOW? The Three Gorges Dam power plant in China can generate enough electricity each day to power 160 million homes!

Renewable Power

The energy for much of our electricity comes from heat made by burning fuels like coal and gas. These fuels create pollution and release carbon dioxide, which damages the climate in very dangerous ways. Renewable power is made using nonpolluting sources of energy, such as sunlight and the wind.

The wing-shaped blades of wind turbines catch the wind and spin around. There is a generator linked to the blades that makes electricity.

Solar Power

Solar energy generates power by capturing light and heat from the sun. The most common form of solar energy uses solar panels, or photovoltaic cells, which convert light energy into electricity. These are often seen on top of houses and other buildings. On a larger scale, solar-thermal power plants can produce power for thousands of people. The sun's energy is used to boil water, producing steam to drive turbines that generate electricity.

Solar panels can be fitted almost anywhere that gets sunshine.

Hydroelectric power plants produce about one seventh of the world's electricity.

Hydroelectricity

This system uses the flow of water to drive an electricity generator. A river is a natural source of flowing water that can be harnessed for hydroelectricity. The power station uses a dam to block the river, and the water is funneled through large tunnels in the dam where its flow makes turbines spin. This spinning motion drives the generators and makes the electricity.

HALL OF FAME: Annie Easley
1933–2011

Annie Easley was a computer scientist and mathematician who worked for NASA. As well as helping design and test space rockets, Easley also began to figure out how the energy in sunlight and the wind could be used to generate power in a renewable way. Toward the end of her working career, Easley devoted a lot of time to helping girls develop an interest in science and engineering and study these subjects at university.

The electric current travels to the land along cables on the seabed.

Wind turbines out at sea work best because they can be made taller, and the wind here blows harder and for longer.

DID YOU KNOW? Solar energy could power the world. The energy in the sunlight that hits Earth each day is about 10,000 times the world's total energy use.

Electronics

Almost every modern electrical appliance, from a washing machine to a phone, is electronic. While electricity works at large scales, with a current driving a motor or lighting a bulb, electronics works at a much smaller scale, with tiny currents and small numbers of electrons. Instead of large circuits, electronics works with tiny circuits, often etched into minute slices of semiconductor.

> This display screen is covered in tiny electronic LEDs. Each LED creates one colored dot, and millions of dots create the full picture.

Transistor

An important electronic device is the transistor. It can work in two ways. It can take in a small current and use it to trigger the flow of a larger current. This is how a hearing aid amplifies a sound, for instance. It can also work like a switch, turning current on and off thousands of times a second. Groups of transistors linked together can form simple decision-making systems that are central to how computers work.

> Transistors work as switches in complex circuits that follow rules set out in a computer program.

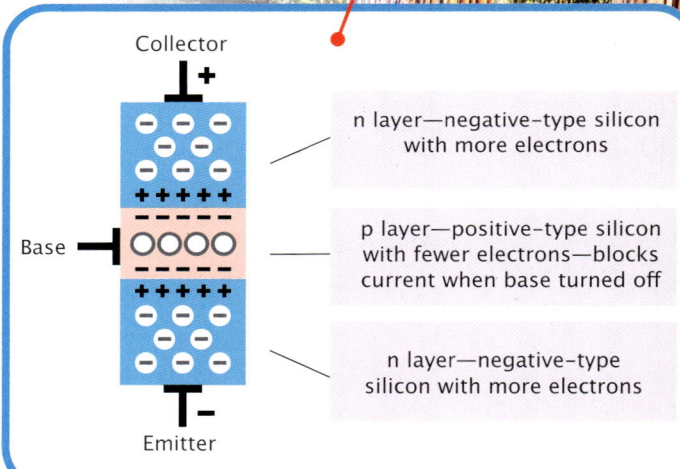

Collector

n layer—negative-type silicon with more electrons

Base

p layer—positive-type silicon with fewer electrons—blocks current when base turned off

n layer—negative-type silicon with more electrons

Emitter

A transistor has at least three connections. Adding a small current to the base connection in the middle causes a larger current to flow between the collector and the emitter.

HALL OF FAME: Esther M. Conwell 1922–2014

Esther M. Conwell was an American physicist who discovered how electrons move through semiconductors. Without this breakthrough, the computers and other gadgets we use would not have been made! A few years after Conwell made her discovery in 1945, the first transistors were constructed. Conwell was the first women to win the Edison Medal, an award that had also been given to Alexander Graham Bell (for inventing the telephone) and Nikola Tesla.

DID YOU KNOW? More than 1 trillion semiconductor components are made each year.

Semiconductors

A semiconductor is a material that has properties somewhere between that of a conductor and an insulator. Most semiconductors are made from silicon, an element found in sand. How well a semiconductor conducts electrify can be changed by adding small amounts of other materials to it. An n-type (n for negative) semiconductor has extra electrons and conducts more easily. A p-type (p for positive) has fewer electrons and is less conductive. The two types can be joined to control the flow of current. Transistors and other components are made from semiconductors.

Silicon is shiny like a metal but it cracks like a crystal.

A diode (including LEDs) is an electronic device that only lets electric current flow through it in one way. It flows from n-type to p-type silicon.

In an LED, electrons move between layers of silicon with different conductivity. As electrons move from n-type to p-type silicon, a tiny burst of light is emitted.

BIOLOGY

Biology is the study of living things. Biologists investigate everything about life, but what makes something alive? There are six features of life. Life can move, reproduce, collect nutrients, use an energy source, sense its surroundings, and grow. Some of these features are seen in nonliving things, like cars or crystals, but only something that has all six is truly alive. There is a lot to discover about living things—from the way tiny microscopic cells work, guided by chemical codes called genes, to how communities of plants, animals, and other types of life rely on one another to survive and make Earth a living world.

Tree of Life

To make sense of the huge variety of life, biologists organize all the different species using a classification system. This system shows how all life on Earth is related, even though evolution has brought millions of species into existence over billions of years.

There are millions of species of animals living across the planet. A species is a group of animals, such as this *Pelagia noctiluca* jellyfish, that look similar and can make babies together.

Plant and Animal Biology

Biologists study how the bodies of plants and animals work—how they get and use energy, how they move, how they reproduce, and how they eliminate waste. Different groups of animals and plants work in different ways.

Cells and Genetics

All life is built from tiny units called cells. Each cell is a microscopic living thing filled with busy chemical machines that carry out the processes of life, known as metabolism. Looking closely at cells can help biologists understand how life as we know it evolved.

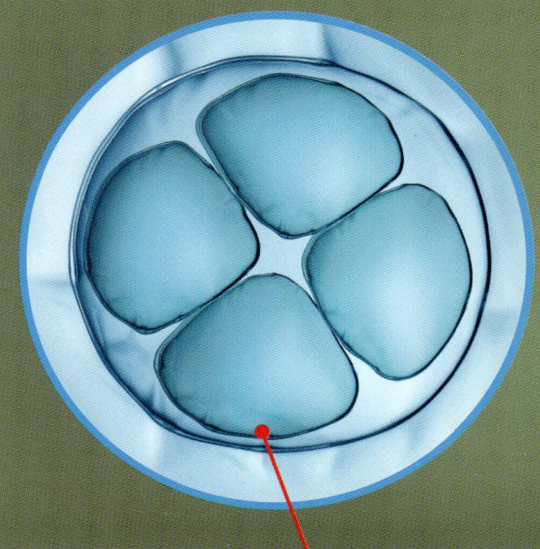

Biologists can use microscopes to look at tiny cells and understand how they work.

Habitats and Ecosystems

Across our planet, the weather and landscapes vary greatly. Yet from the frozen poles to the hot rain forests of the equator, living things make their homes. Different species have adapted over time to different temperatures and terrains. Biologists study how organisms relate to one another and their environment.

This panther chameleon lives only in Madagascar in a tropical forest biome.

The human body is a collection of systems that each perform a set of vital functions that keep us alive.

Chapter 8: Tree of Life
Classification

More than one million species, or types, of living thing have been described by biologists so far, and scientists predict that there are many millions more to be discovered. To make sense of the great variety of life, all organisms are organized into groups using a system called classification.

Connecting With the Ancestors

The aim of taxonomy is to organize all life on Earth according to how the different groups are related to each other. All members of a particular taxon share a common ancestor. That means that all members of a class or order evolved from one species a long time ago. A small group, like a genus, has only a few species, and their common ancestor probably lived quite recently. A bigger group like a phylum has hundreds of thousands of species, and their single common ancestor lived way, way back. The first animal, for example, was a tiny wormlike creature from more than half a billion years ago.

A type of feathered dinosaur is the ancestor of today's birds.

Binomial System

Scientists have given every species a scientific name made of two words. The first word is the generic name, or the name of the genus the species belong to. The second word is the specific name, which is a unique name for that species. This binomial system was created to prevent confusion.

Many of the words used in taxonomy are based on old languages such as Latin and ancient Greek. They often describe the group in some way. For example, bats belong to an order called Chiroptera, which means "hand wings." Our own species name, Homo sapiens, means "wise human."

The science of classification is called taxonomy. It divides life into a series of groups called taxons, which are ranked by size. The main basic taxon is the species. A species is a group of living things that can breed and produce young together. For example, all humans belong to one species called *Homo sapiens*.

The human species is part of a larger grouping known as a genus, called *Homo*. There used to be other species in this genus, such as *Homo erectus* and *Homo neanderthalenis*, but they have died out.

The *Homo* genus belongs to a family, Hominidae, that it shares with other great apes: chimpanzees, gorillas, and orangutans.

The hominid family is one of several in the Primate order, which also includes monkeys, lemurs, and gibbons.

Primates are members of the Mammalia class, along with whales, lions, and mice.

The mammals are one class of the phylum Chordata, which also include reptiles, birds, and fish—anything with a backbone.

All the animal phyla combine to make the Animalia, or animal kingdom. There are several other kingdoms of life on Earth, including plants, fungi, and bacteria.

TAXON	HUMAN	CHIMPANZEE	BLUE WHALE	SNAKE
Species	*sapiens*	*troglodytes*	*musculus*	*naja*
Genus	*Homo*	*Pan*	*Balaenoptera*	*Naja*
Family	Hominidae	Hominidae	Balaenopteridae	Elapidae
Order	Primates	Primates	Artiodactyla	Squamata
Class	Mammalia	Mammalia	Mammalia	Reptilia
Phylum	Chordata	Chordata	Chordata	Chordata
Kingdom	Animalia	Animalia	Animalia	Animalia

HALL OF FAME: Carl Linnaeus
1707–1778

The taxonomy system used today was originally organized by Swedish plant scientist Carl Linnaeus. He set up the system in 1735. It had the same set of taxons as today, but Linnaeus did not understand the process of evolution. Instead he grouped organisms according to how they looked. This led to early mistakes, such as grouping whales and dolphins as kinds of fish, though he later changed this.

DID YOU KNOW? Taxonomists have shown that the fungal kingdom is more closely related to the animal kingdom than it is to the plant kingdom.

Bacteria

Among the smallest and oldest type of life are bacteria. Fossil remains show that bacteria were living on Earth at least 3.5 billion years ago. Bacteria are microscopic organisms and far too small to see without powerful microscopes. They live in all habitats, from the deep ocean to the ice at the top of mountains. They even live in rocks deep underground and float around in the air.

Bacteria evolved at a time when the conditions on Earth were much more extreme than they are now. As a result, bacteria can be found living in places, such as this hot spring, where no other life form can survive. Bacteria also survive in ice and in toxic chemicals.

Single Cells

A single bacterium has a body made of one cell, which is about 2 micrometers long (that's 2 millionths of a meter). The cell is surrounded by a membrane and a rigid cell wall. Inside the cell is held a complex mixture of DNA and other chemicals of life. Most bacteria cells are spherical or rod-shaped but a few have a twisted shape. They sometimes grow into chains or clusters made up of several cells.

One of the most common types of bacteria are the cyanobacteria (also called blue-green algae). Vast numbers of them float in the ocean as plankton. One of the most common species on Earth is *Pelagibacter*. It is estimated that for every human on Earth, there are 3 million trillion of these cyanobacteria in the sea.

HALL OF FAME: Ruth Ella Moore
1903–1994

In 1933, Ruth Ella Moore became the first Black American woman to earn a PhD in biology. She was an expert in bacteria and worked on the germs that caused tuberculosis—a lung disease that is still one of the biggest killers today. Moore also showed that gum disease and rotting teeth were caused by bacteria in the mouth that lived on sugary foods.

The colors in the water come from the microbes living there and their waste chemicals.

Bacteria, Good and Bad

Some bacteria cause illness such as an upset stomach and sore throat. They can be treated with germ-killing drugs called antibiotics. Bacteria also infect wounds and damage the body, and so they must be cleaned away with antiseptics. However, bacteria are also in foods such as yogurts and pickles. In fact, the bacteria make these foods taste the way they do by adding acids. These food bacteria are useful to the body because they help with digestion. There are billions of bacteria living inside your intestines right now.

Another group of microbes called archaea is found in these hot waters. Archaea look a lot like bacteria and have been around just as long. However, they have a distinct metabolic system, and so form a separate kingdom.

Yogurt is made when bacteria turn the sugars in milk into lactic acid. This gives the sharp taste and creates the gooey mixture.

DID YOU KNOW? The weight of all bacteria on Earth is around 70 billion tonnes (77 billion tons)—45 times heavier than all the world's animals combined!

Protists

The protists are single-celled organisms that have much larger and more complicated cells than bacteria and archaea. Some biologists put all the protists into one kingdom, but they are a varied group of organisms. The cells of protists have many internal structures, just like the cells of multicellular life-forms, such as plants and animals. Some protists live like animals, and others are more plantlike. Some are like both at the same time!

Flagellates and Ciliates

A large group of protists get their name from extensions on their cells called flagella and cilia, which help the cells to move. Flagellates are a common type of plankton and are responsible for "algal blooms" where seawater is filled with these microbes, choking out other forms of life. Ciliates also live as plankton but are common in soils and even live as parasites inside animals. They waft their cilia to pull tiny particles of food to a mouthlike opening in the cell.

These flagellates have one flagella each, but some have two, three, or several dozen.

Diatoms

These common types of plantlike protists live in seawater and in lakes and rivers. Some species live in damp soil. Diatoms use photosynthesis to make a supply of food from sunlight. They are either rounded or boat-shaped, and the cell sits inside a shelllike case made from silica (the same chemical in sand).

Some animallike protists are called amoebas. The cell has no rigid walls, and so it can squirm into any shape. Many amoebas are parasites that live inside animal bodies, often causing diseases.

The diatom's case is in two halves. The lower half always fits snugly inside the upper one.

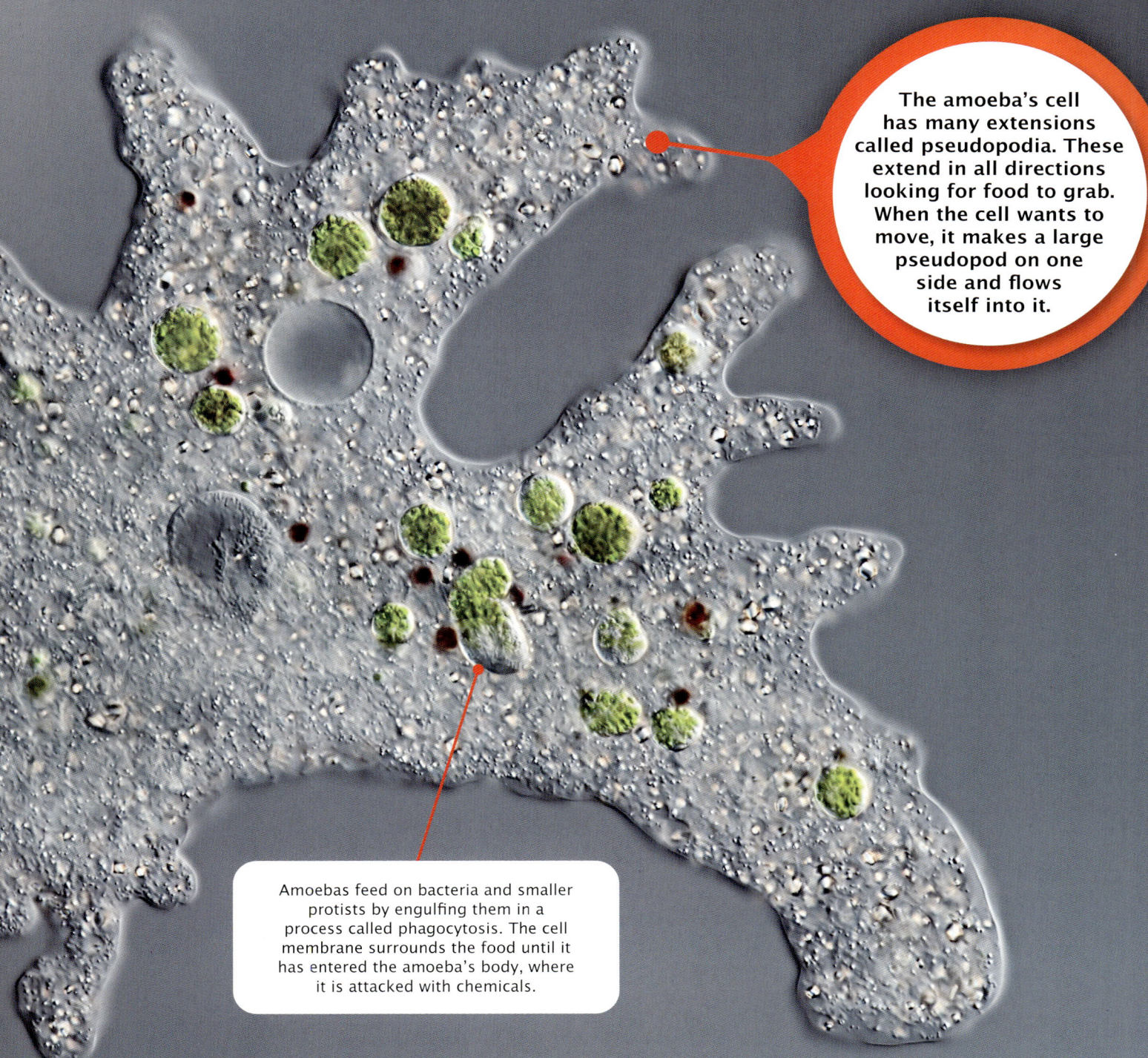

HALL OF FAME: Antonie van Leeuwenhoek
1632–1723

The first person to see protists was Antonie van Leeuwenhoek. This Dutch businessman made much improved versions of the microscope in the 1670s. He discovered a hidden world of microscopic life and called the organisms he saw "animalcules," meaning "little animals." He described many types that we now recognize as ciliates, amoebas, and other kinds of protists.

The amoeba's cell has many extensions called pseudopodia. These extend in all directions looking for food to grab. When the cell wants to move, it makes a large pseudopod on one side and flows itself into it.

Amoebas feed on bacteria and smaller protists by engulfing them in a process called phagocytosis. The cell membrane surrounds the food until it has entered the amoeba's body, where it is attacked with chemicals.

DID YOU KNOW? A protist called *Plasmodium* causes the deadly disease malaria. Each year a quarter of a billion people catch it and around 600,000 die. A new vaccine is being tested to stop *Plasmodium*.

Plants

There are around a quarter of a million species of plants, making up the kingdom Plantae. Members of this kingdom range from tiny mosses to towering trees. Plants power their bodies using a process called photosynthesis. This uses the energy in sunlight to make sugar from water and carbon dioxide. Plants are found in all parts of the world except the coldest and driest habitats.

The giant sequoia is one of the largest and longest-living organisms on Earth. The tree from western North America grows to about 85 m (280 ft) tall and lives for more than 3,000 years!

Internal Vessels

Moss is one of the simplest kinds of plants. It grows over surfaces, and its flat body has no distinct roots, stem, or leaves. Mosses, along with liverworts, are nonvascular plants. Most plants are vascular, which means they have a network of vessels running inside the body. They include ferns, conifers, and flowering plants. The vessels transport water and sugar around the plant. They also make the plant's body stiff enough to grow up toward the light.

The giant sequoia is a conifer. It uses cones to breed and make seeds. Most plants grow flowers, not cones, for this purpose.

Plants like this moss are mostly green because they use a green chemical called chlorophyll to absorb red and blue sunlight and use its energy to make food.

HALL OF FAME: Janaki Ammal 1897–1984

Born in India, Janaki Ammal studied to be a botanist, or an expert in plants. She was one of the first women to do this. Ammal used her knowledge to breed new kinds of crops that would grow better in India and allow the country to produce its own food. However, she also campaigned to keep as many of India's natural habitats as possible.

Seaweed

The "plants" that live in the oceans are called seaweeds. They are not normally included in the Plantae kingdom. Instead they are algae, types of protists, that grow into large, multicellular bodies. Seaweeds photosynthesize and need sunlight to survive like land plants do, and so they grow mostly in shallow, sunlit water. There are red and brown types, as well as green. Seaweeds have no roots but are anchored to the seabed. Instead of leaves, they have fronds that float in the water.

Seaweeds are exposed to the air when the tide goes out. To stay moist until the sea rises again, many seaweeds cover their fronds in waterproof slime.

Trees use their height to grow above smaller plants so they can collect more sunlight. To grow tall, the trees strengthen their bodies with wood. This is a hardened material made from the internal system of vessels running up inside the trunk.

DID YOU KNOW? Plants make up 80 percent of all the living material on Earth! In total that is 450 billion tonnes (496 billion tons).

Fungi

The Fungi is a third kingdom of multicellular organisms alongside the plants and animals. There are 140,000 known species of fungi. The most familiar are the mushrooms and toadstools that sprout from the ground, but these are only the fruiting bodies that grow to spread the fungus's spores in the water and wind. Most of the fungus is growing unseen below the ground or inside plants—and even on our bodies!

The bright colors of this toadstool are a warning to animals not to eat the fungus. It contains dangerous poisons. Most mushrooms and toadstools will make you sick, so only eat the ones sold in stores.

External Digestion

Fungi are saprophytes, which means they grow on their food. They do not have a mouth or stomach. Instead they release digestive enzymes that break food down into a sloppy mush outside the fungus's body. Then the fungus absorbs the useful nutrients. Fungi are very important members of the natural habitats because they drive the decay of dead material, eating it up and recycling important nutrients into the soil.

This blue-green mold is a kind of fungus. It thrives in damp conditions and grows from microscopic spores that float in the air and land on leftover foods.

Food and Fungus

Many mushrooms are edible and contain minerals and vitamins. Fungi are ingredients in other foods, as well. The blue part of blue cheese is a fungus, and most bread is made with a microscopic fungus called yeast. The yeast eats the sugars in the dough, releasing carbon dioxide gas. Bubbles of this gas make the dough rise and create a soft, springy loaf.

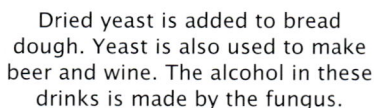

Dried yeast is added to bread dough. Yeast is also used to make beer and wine. The alcohol in these drinks is made by the fungus.

DID YOU KNOW? The largest living thing on Earth is not a whale or a tree. It is a honey fungus living in the soil of Oregon. The fungus covers the area of 1,665 soccer fields!

The main body of a fungus is made up of many thin filaments called hyphae. These have stiff walls made from chitin, which is the same chemical used to make insect bodies and crab shells. The network of hyphae is called the mycelium.

Toadstools grow out of the fungus's main body, or mycelium, which is hidden in soil or dead wood. The cap of the toadstool opens up, and microscopic spores are blown away in the wind. New mycelia will grow from the spores.

Toadstools can grow very fast, almost overnight. When the conditions are right, the cells in the stalk will grow longer and longer. This pushes the cap up out of the ground.

HALL OF FAME: Alexander Fleming
1881–1955

Alexander Fleming, a Scottish microbiologist, made the most important fungus discovery in history. In 1928, mold contaminated his laboratory samples of bacteria. He noticed that the fungus had killed the bacteria around it. The fungus was making an antibiotic chemical, which was later named penicillin. Penicillin and other antibiotics are used today to stop dangerous infections. They have saved hundreds of millions of lives!

Simple Invertebrates

Around 97 percent of animals are invertebrates. An invertebrate is any animal that has no backbone or hard skeleton inside the body (unlike vertebrate animals, such as humans). Instead invertebrates come in a great variety of forms. The simplest of all are the sponges, which form funnel-shaped bodies for filtering food from the water. Other invertebrates include worms, jellyfish, and mollusks.

Soft Bodies

The jellyfish belong to a larger group of animals called the cnidarians. They are all soft-bodied animals with tentacles. The tentacles have stinger cells that fire poison darts into anything that touches them. Cnidarians all live in water, and they also include corals and sea anemones. Unlike most other animals, cnidarians do not have a head. Instead their bodies are rounded with a mouth in the middle.

A sea slug, or nudibranch, is a kind of mollusk that has no shell. Other mollusks without shells include squids and octopuses.

This species is called the fried egg jellyfish. A large jellyfish like this swims by pumping out jets of water by squeezing its bell-shaped body.

HALL OF FAME: Hope Black
1919–2018

Hope Black was a leading Australian malacologist, or expert on mollusks. She started work at National Museum of Victoria while still a teenager. Within ten years, Black was the first woman to be made a national curator in Australia. In 1959, Black was on one of the first female teams to explore Australia's Antarctic islands.

Mollusks

Mollusks are a large group of invertebrates. Many mollusks, such as snails and slugs, live on land, but most them are aquatic animals. Most mollusks protect their body with a shell. Snails have a single shell, while shellfish like clams and oysters have two shells connected by a hinge.

The mollusk shell is made mostly from a hard stonelike chemical called calcium carbonate. This land snail uses the shell for protection and to keep its body moist.

Like most animals, the sea slug has bilateral body symmetry, which means the right and left halves of its body are mirror images. There is a head at one end where the mouth, brain, and main sense organs are located, and a rear opening for waste at the other end.

The antennae on the head of the sea slug are rhinophores that are used for picking up chemicals in the water.

DID YOU KNOW? A large sea snail called the geography cone shell is one of the most venomous animals in the world. Its venom is 10 times more powerful than a king cobra's.

Arthropods

The largest animal phylum is the invertebrate group called arthropods. That name means "jointed foot," and it refers to how these animals have an armorlike exoskeleton, or hard outer skeleton, made up of interlocking jointed sections. The exoskeleton is made from chitin, a flexible plasticlike material. The arthropods include three main subgroups: the insects, the arachnids, such as spiders and scorpions, and the crustaceans, such as crabs and prawns.

Insects

The insects are by far the largest group of arthropods. In fact, 80 percent of all animal species are insects. Insects have six legs and bodies in three sections: the head, thorax, and abdomen. Often there are one or two pairs of wings on the thorax. It is thought that insects were the first animals to evolve flight around 320 million years ago. Familiar kinds of insects include beetles, flies, ants, and butterflies.

Like all beetles, this scarab has a tough cover protecting its wings.

Crustaceans

Crustaceans have a varied number of legs, and often limblike appendages are used as pincers. Most of these animals live in the ocean. Copepods and krill, two types of crustaceans that live as plankton, are among the most numerous animals on Earth. Larger crustaceans like lobsters toughen their exoskeleton with calcium carbonate. Other crustaceans include barnacles that glue themselves to a rock and filter food from the water with their feathery legs.

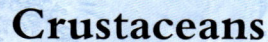

Wood lice, also called pill bugs, are among the few crustaceans that live on land. However, they only survive in moist habitats such as among fallen leaves. Some species can roll up into a ball to fend off threats.

DID YOU KNOW? Very few kinds of insects live in the sea. The main group is called the sea skaters. All others live on land or in fresh water.

The arachnids have eight legs and limblike mouthparts called palps. All arachnids are hunters, and most kill using venom. The venom turns the prey to mush which the arachnid then slurps up through their small mouth.

Like all arachnids, this spider has two body sections. The legs are attached to the forward section, called the cephalothorax. The rear section is the abdomen. Unlike other arthropods, arachnids have no antennae or feelers on the head.

Spiders use silk to catch prey. This one has created a netlike web of sticky silk that snares flying insects. The spider uses its feet to pick up any vibrations made by prey or other spiders.

HALL OF FAME: Margaret S. Collins 1922–1996

Margaret S. Collins was such a clever child that she was allowed to start attending university at the age of 14. She became only the third Black American woman to become a doctor of zoology. Collins became an expert in termites—wood-eating insects that live in huge colonies. Termites can damage houses made from wood, and Collins wanted to learn as much about them as she could. Collins was also a leading figure in the Civil Rights Movement in Florida.

Lower Vertebrates

> Amphibians spend the first stage of their life in water swimming around as fishlike tadpoles. They then grow legs and transform into adults that move out onto the land.

The first vertebrates were fish that evolved around 500 million years ago. All of today's vertebrates (animals with backbones and internal skeletons) evolved from fish. The first land vertebrates were amphibians—the ancestors of today's frogs and salamanders. These land animals evolved into reptiles and the ancestors of mammals. Birds evolved later from dinosaurs (a type of reptile).

Reptiles

The reptiles are a varied group. They all have skin covered in tough waterproof scales, and they are not reliant on water to breed (unlike fish and amphibians). There are three major types: the turtles and tortoises, the crocodiles, and the squamates. This last group is by far the largest and contains snakes and lizards. Most reptiles lay eggs with a waterproof shell, but a few give birth to their young. They are ectotherms, or cold-blooded, meaning their bodies are the same temperature as the surroundings.

Snakes have evolved to slither on their bellies without legs. There are about 4,000 species, and 600 of them, including this blue viper, use venom to kill prey.

HALL OF FAME:
Bertha Lutz
1894–1976

Born in Brazil and educated in Paris, France, Bertha Lutz became a leading expert in poison dart frogs. These brightly colored little amphibians collect poisons from the ants and other insects that they eat and store them in their waxy skin. The poisoned skin makes any predators sick—or even die—so they learn not to attack the frogs. Alongside her career as a zoologist, Lutz helped ensure Brazilian women were allowed to vote and was also involved in setting up the United Nations. Two frog species and four lizards are named after her.

The gills are located behind the head. Water enters the mouth, flows through the gills (where oxygen is transferred to the blood), and then out through slits on the side of the neck.

Fish

There are 33,000 species of fish. They live in the oceans and in freshwater rivers and lakes. They use gills to take oxygen from the water, although a few species can breathe air for short periods. Fish have a basic streamlined body that flows easily through the water. There is a tail fin for swimming and a number of side fins for steering. The back, or dorsal, fin stops the fish from rolling onto its side as it swims.

All frogs are hunters. The horned frogs of South America have mouths that are large enough to swallow prey that is the same size as the frog!

Adult frogs have no tails. They move by hopping, while salamanders and newts keep their tails into adulthood. Amphibians return to water to breed. Their eggs have no shells and will dry out unless they are laid on or under the water.

DID YOU KNOW? The marine iguana is the only lizard to feed under the sea—it eats seaweed. When food is scarce, the lizard shrinks in size to save energy.

Birds

There are around 10,000 species of birds, most of which are capable of flight. All of them have two legs and a pair of wings. To take to the air, a bird's body must balance low weight with great strength. Flightless birds, such as ostriches and penguins, do not have this issue. The ostrich has swapped flight for a large size and great running ability, while penguins use their wings as flippers for swimming, not flying.

Birds have no teeth. Instead the mouth is formed by a hard beak or bill. The shape of the beak indicates what food the bird eats. Hooked beaks are for ripping and cracking foods. A pointed beak is best suited for picking up small items, like insects.

Primitive Birds

The birds evolved from dinosaurs around 150 million years ago. The descendants of these first birds are known as fowl and include waterbirds, like ducks and geese, and ground birds, like chickens and this partridge (right). Waterfowl are strong fliers that often make long migrations by air. The mute swan is one of the largest flying birds of all. However, ground birds spend most of their time feeding on the ground and are only capable of short fluttering flights to escape danger.

The shape of a bird's wing shows how they fly. The square wing of this eagle is ideal for soaring and slow, controlled flight. Smaller triangular wings are for faster flight with tighter turns.

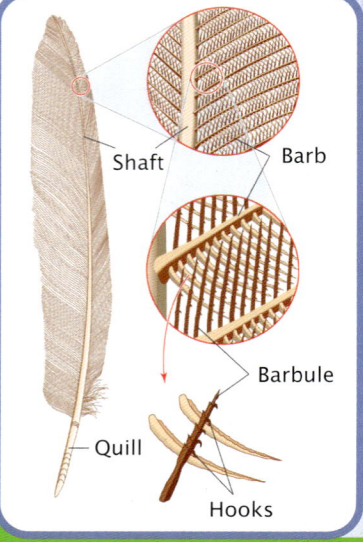

Shaft · Barb · Barbule · Quill · Hooks

Feathers

The first animals to have feathers were dinosaurs. These animals used them for warmth. Birds do this, too. The feathers near the skin are small and fluffy and trap a blanket of warm air. Feathers are made from keratin, the same material in mammal hair and reptile scales. Wing and tail feathers are flat and stiff because the many branches of keratin are neatly linked together.

Bird skeletons have no tail section. However, the birds create a tail with long feathers. The tail helps with flying. It is used for steering and braking. Some birds use their long tails to communicate and get attention.

Songbirds, like this multicolored tanager from South America, often have brightly colored feathers. This is to make it easier for mates to find each other. Many birds also sing regularly to advertise their presence.

HALL OF FAME:
John James Audubon
1785–1851

This French–American artist and naturalist is famous for making an extensive record of all North American birds and painting each species. John James Audubon published his full set of pictures from 1827–1838, and they are still a good way of identifying birds. The National Audubon Society was set up in his name to protect birds in North America and across the world.

DID YOU KNOW? The hooded pitohui from New Guinea is one of the world's only poisonous birds. It collects chemicals from its ant food and stores them in the skin.

Mammals

The most varied and widespread vertebrates are mammals. All mammals have hairy bodies for at least some of their lives, and they all feed their young on milk. Beyond that, they come in all shapes and sizes. The smallest mammals are the size of a thumb, while the largest—the 24-m (79-ft) blue whale—is as long as two buses. Mammals are warm-blooded, which means they use energy from food to maintain a constant body temperature. As a result, mammals can survive everywhere from the icy polar seas and high mountains to steamy jungles and dry deserts.

Marine Mammals

Some mammals have returned to live in the oceans. They have lost their legs and can no longer walk on land. Instead their limbs are flippers, which are much more useful in the water. Marine mammals include the cetaceans, such as dolphins and whales, which have lost their back legs entirely and never come on land. There are also the pinnipeds, better known as the seals and sea lions, which return to beaches to rest and are able to shuffle short distances on land.

This antelope calf is an example of a hoofed mammal. These plant-eating mammals have long legs tipped with tough hooves, which are the equivalent of very thick toenails. Thanks to their long legs, most hoofed animals are fast runners.

Like all cetaceans, dolphins have smooth skin (they have hairs before they are born) and a nostrillike blowhole on top of the head for breathing at the surface.

Hair is made from shafts of dead cells coated in keratin. It grows from a root embedded in the skin. Keratin is also used to make fingernails, claws, and hooves. Some animals have a coating of keratin to make the skin waterproof.

DID YOU KNOW? The musk ox is the mammal with the longest hairs. Some of the hairs grow to 1 m (40 in) long. The musk ox lives in the Arctic. Its long hair forms a thick curtain against the cold wind.

Marsupials

Most mammals are born after developing for some time inside their mother's uterus (womb). Female kangaroos and other marsupial mothers have a small uterus. Their babies are born early in their development, when very small. The baby, or joey, then moves to a pouch on the mother's belly, where it drinks milk and continues its development. Marsupials are the main kind of mammal in Australia, although there are marsupials living in North America and South America, too.

Kangaroos do not walk. Instead they hop along on their large, bouncy back feet. This is an efficient way to get around, especially when carrying a big joey in the pouch.

The cheetah is the fastest running animal of all. It can reach speeds of 110 km/h (68 mph), but only for a few seconds. All that exertion makes the cat very hot, and it has to stop and cool down.

HALL OF FAME:
Jane Goodall
1934–

When she was 26 years old, Jane Goodall went to live alongside a group of chimpanzees in the forests of Tanzania. She watched how the apes behaved and communicated, and she built up a picture of how chimp society worked. One discovery Goodall made was that chimpanzees constructed simple tools to collect food. Ever since, Goodall has been studying primates and calling for wild places to be better protected.

Chapter 9: Plant and Animal Biology

Photosynthesis and Respiration

All life on Earth requires a source of energy to power its metabolism, or life processes. The energy used by life is delivered by two processes, photosynthesis and respiration. Photosynthesis is used by plants to convert the energy in sunlight into sugar fuels. Respiration releases the energy in sugars for use in metabolism.

Chloroplasts

Photosynthesis occurs in plant cells inside green capsules called chloroplasts. They are green because they contain the pigment chlorophyll. When sunlight hits the chlorophyll, some of its energy is absorbed. This energy is used to react carbon dioxide gas from the air with water to make a simple sugar called glucose. The reaction produces oxygen as a waste product, and this is given out by the plant. The same process of photosynthesis occurs in some bacteria, but bacteria do not use chloroplasts to hold chlorophyll.

Membrane stack

Chlorophyll is bonded to stacks of membranes inside the chloroplast. This molecule looks green because it absorbs red and blue light but reflects back the green beams.

Mitochondria

Respiration is more or less the reverse of photosynthesis where oxygen reacts with glucose to release energy as it breaks up into carbon dioxide and water. In complex life, respiration happens inside structures in the cell called mitochondria. The glucose and oxygen react in several small steps so energy is released slowly. The carbon dioxide produced by respiration must be removed from animal bodies, although plants can use it again for photosynthesis.

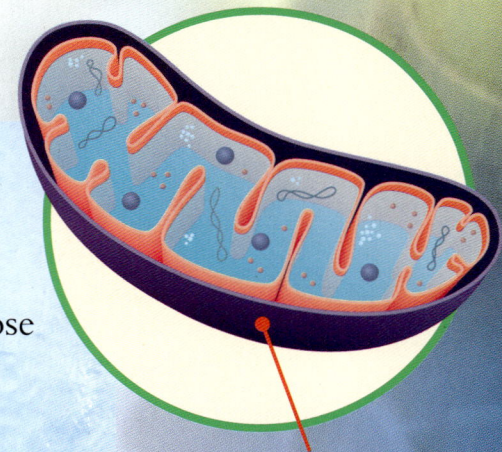

Mitochondria are surrounded by two membranes. Respiration happens close to the inner one, which is highly folded to increase its area.

Plants are autotrophs, or "self eaters," because they use photosynthesis to create their own food supply. They use this food in respiration whenever needed.

Animals are heterotrophs, or "other eaters." They survive by eating the body parts of other organisms, using respiration to release energy from this food.

This hummingbird is using a lot of energy to hover as it feeds. Muscle cells have many thousands of mitochondria to release the energy they need.

HALL OF FAME:
Jan Ingenhousz
1730–1799

Photosynthesis was discovered by this Dutch biologist in 1779. At this time, scientists were only just beginning to understand that air was filled with separate gases, such as oxygen and carbon dioxide. They also knew that plants gave out these gases in different conditions. Jan Ingenhousz showed that oxygen is made when a plant is in sunlight, but that this stops when it gets dark.

DID YOU KNOW? Before photosynthesis evolved around 3.5 billion years ago, there was no oxygen in the air. In fact, oxygen was poisonous to most living things back then.

Plant Bodies

Plants range in size from the mightiest tree to the tiniest daisy, but they follow the same basic body plan. The lower part is made of roots that extend into the soil. The middle of the plant is a stem that often divides into branches before sprouting leaves on the upper part of the plant.

Transport Vessels

Most plants have an internal network of vessels that run up the stem between the roots and leaves. Xylem tissue carries water and dissolved nutrients and salts. It is made from dead cells with open ends, and so forms stiff tubes that also provide structural support to the plant. Phloem tissue transports sugars, the plant's energy source, from the leaves where they are made to the rest of the plant.

The older xylem tubes in thicker stems are filled in with a hard material to make wood. Woody stems, or trunks, are very strong, and so plants can grow to heights of more than 100 m (330 ft).

Water always flows upward along xylem, while the sugars in phloem can travel in either direction.

Leaf Anatomy

A leaf is a plant's solar panel and is thin and flat, so that it catches as much light as possible. The light enters through the top, and here the cells are packed with chloroplasts to capture its energy. The water needed for photosynthesis is provided by vessels in the leaf's central vein. Carbon dioxide from the air enters gas spaces in the lower part of the leaf through pores called stomata.

The stomata are mostly on the underside of the leaf to prevent water from evaporating away rapidly in strong sunlight.

DID YOU KNOW? Pando is a forest of about 40,000 aspen trees in the United States that are all connected underground, making it the world's largest plant.

HALL OF FAME:
Agnes Arber
1879–1960

Agnes Arber became interested in botany, or the science of plants, while still at school. After a long education, she became a full-time scientist and started to publish books on plants. These works were groundbreaking in the way they described plant bodies scientifically. Arber was an expert in conifers and grasses.

Roots need a supply of air as well as water. These mangrove trees have sturdy woody roots above the water, so they can access the air.

Bark is not quite the same as wood. It has waxy chemicals in it to make it waterproof, and there are often smelly oils and resins that ward off insects and other invaders.

Plant Reproduction

The two main types of plants that grow on land—the flowering plants and conifers—reproduce by making seeds. Seeds are created when pollen grains carrying male sex cells fuse with female ovules (which hold the egg cells). During this process, known as pollination, pollen is moved from plant to plant by the wind, water, or animals.

Flowers use bright colors and powerful odors to attract insects and other animals. The animals come to eat nectar, a sweet liquid produced the by the flower.

Flowers

The flower is a plant's sexual organ. Pollen is produced on tall stalks called stamens, while the ovules are held in the ovary in the center of the flower. Wind-pollinated flowers produce dry, dustlike pollen that blows away easily. Flowers that rely on animals, such as insects, for pollination create sticky grains that cling to animal bodies. Pollen from another flower sticks to the central stigma and burrows into the ovary.

Stigma, Stamen, Ovary, Petals

After pollination, the ovary develops seeds, and the surrounding region grows into a fruit of some kind. The fruit is designed to spread the seeds somewhere they can grow.

HALL OF FAME: Karl von Frisch
1886–1982

Honeybees are famous for dancing. The dance is a way for one honeybee to communicate to others the direction and distance to a good foraging location full of flowers. Karl von Frisch, an Austrian biologist, discovered the honeybee dance and translated what it means. He won the Nobel Prize in 1973 for this important breakthrough.

DID YOU KNOW? The smallest pollen is made by the forget-me-not flower. It is about 6 microns (0.006 mm), about the same size as a bacterium.

Germination

The process by which a seed sprouts into a new plant is called germination. Germination is stimulated by temperature, water levels, and the length of daylight. When conditions are right, a shoot will emerge from the seed and grow toward the light and away from gravity. The seed contains one or two embryonic leaves called cotyledons, which contain a store of nutrients that fuels growth until the first true leaves can form and start to photosynthesize.

The plant's first root, the radicle, does the opposite of the stem. It grows toward the pull of gravity and away from light.

Honeybees are one of the most important pollinators. The bees carry nectar and pollen back to the nest. The nectar and pollen are the raw ingredients for making honey, the bees' main food.

Animals carry pollen grains from flower to flower, where they can be used in reproduction. Some of the animals will eat the pollen as well, but there is plenty to spare.

Animal Bodies

There are thought to be several million animal species, although most are still unknown to science. Each one has evolved to live in a unique way in a specific habitat, and so there is huge diversity in their shape and size. One thing animals have in common is a body that is capable of actively moving to gather food.

Fish have a solid internal framework, or skeleton, made from bone and cartilage. This gives the animal its shape.

Body Symmetry

Most animals have a bilateral body symmetry, which means that the left side of the body is a mirror image of the right. This bilateral plan is seen in everything from worms to whales. Bilateral animals have a head at one end, a waste opening at the other, and a distinct top and bottom (the belly). However, a few primitive animals, such as the sponges, have bodies with no head and no symmetry at all. Jellyfish and their relatives have round bodies.

This centipede's body is constructed from several repeating segments. Many animals have a body organized like this, with different segments specializing for different tasks.

Frequent Changes

Animal bodies can change a lot through the life cycle of species. All kinds of animals, from insects to frogs, have a larval phase, where the young animals look very different to the adults and survive in a completely different way. For example, a caterpillar larva eats leaves while an adult butterfly sips nectar. Similarly, mammals and birds will develop thicker coats of hair or feathers in winter and shed them for a cooler covering in summer.

This snow monkey keeps warm on cold winter days by taking a bath in warm, volcanic springs. Its fur has also become much thicker but will thin out in spring.

HALL OF FAME: Pierre Belon
1517–1564

This French zoologist was one of the first scientists to study comparative anatomy. Pierre Belon compared the body structures of different animals, such as humans and birds, to look for what made them different and which body parts were the same. This process was an important step in understanding how animals—and all other kinds of life—are always evolving.

Sea anemones are named after a type of plant (anemones are pretty flowers). Early biologists thought they were seaweeds, but sea anemones are all meat-eating killers!

The sea anemone has no hard body parts. Instead it has a hydrostatic skeleton made up of capsules of water. These capsules are flexible but retain their shape. This gives the anemone's muscles something to pull against.

DID YOU KNOW? The longest animal in the world is not the blue whale (that's the largest) but the siphonophore. These close cousins of jellyfish have tentacles 45 m (150 ft) long, almost twice the length of a whale!

Animal Locomotion

There are several forms of animal locomotion—the most primitive is swimming. This generally involves the movement of the whole body to push the animal through the water. Some animals, such as squid, propel themselves with jets of water. Walking, running, and climbing all make use of legs and feet, but many land animals get around without them.

Moving by jumping is called saltation. Kangaroos are famous for moving like this. The animal leans forward and raises its long tail to stay balanced in the air.

Air Time

Animal movement over land often involves spending time off the ground in the air. During walking, at least one foot is on the ground at any time. Running is faster, so all the animal's feet are off the ground for short leaps. Gliding is a not quite the same as flying. It is a slow, controlled fall from a high start point to a lower landing site. Expert gliders, such as sugar gliders, can stay in the air for several seconds.

True flight is when the wings create a life force that allows animals to travel up away from the ground. Only four types of animals have evolved true flight: insects, bats, birds, and pterosaurs. (This last group became extinct along with the dinosaurs.)

HALL OF FAME:
Eadweard Muybridge
1830–1904

Eadweard Muybridge was a pioneer of moving images. He used many cameras set out in a line to take photographs in order. Together, they could be made into an early form of video. Muybridge used his system to capture the way animals, mostly horses, moved. His videos showed how the legs moved differently when the animal walked, trotted, and galloped.

Slithering

The snake is a familiar example of a legless animal, but others include caterpillars, worms, and maggots. There are three main ways a legless animal can slide. In rectilinear motion, the body ripples in an up-down wave as it slides forward. In the second method, the body loops from side to side as it pushes on the ground. The third method is sidewinding, where the wavelike body movements push the animal sideways instead of forward.

Sidewinding is useful for moving efficiently over loose ground like desert sand.

The forelimbs are free for holding food, cleaning fur, and boxing rivals!

Kangaroos are known as macropods, meaning "big feet." The tendons connecting the feet to the leg muscles are very stretchy, so the animal bounces without having to use much energy.

DID YOU KNOW? Young spiders create sails of electrified silk strands to catch the wind and take to the air. They can travel long distances using this method, which is known as ballooning.

Animal Reproduction

The aim of life is not just to survive but to reproduce and increase in number. Animals will divert their resources to reproduction, often going without food in order to succeed. Sexual reproduction is the most common method for animals, where each new individual has two parents, but there are nevertheless many different strategies to get the best results.

This aphid is giving birth to daughters using asexual reproduction. The baby insects have no father, only a mother, so she can produce offspring very quickly without needing to find a mate.

Aiming for Quantity

Smaller animals often devote their resources to producing large numbers of young. The females produce large numbers of small eggs, and the males generally fertilize them after they are released. The parents have little ability to take care of large numbers of young, and inevitably many will die before they are able to breed themselves. However, this strategy allows just a few animals to rapidly populate a new habitat if that opportunity were to appear.

Frog spawn is left unguarded in the water. The eggs will hatch into tadpoles that must fend for themselves, and most will not make it to adulthood.

HALL OF FAME: Aristotle 384–322 BCE

The way animals reproduce was not fully understood until the early twentieth century, when the science of genetics and inheritance showed how DNA was passed from parents to offspring. Long before that, people believed that small animals, like worms and aphids, emerge spontaneously from rotting material. This idea was first put forward by Aristotle, a Greek philosopher, who some say was one of the first biologists. He made detailed records of sea life in the waters near his home.

DID YOU KNOW? The oceanic sunfish produces 300 million eggs each year. Only a tiny fraction—perhaps one or two—of these eggs will reach full adult size.

Orangutans have the longest childhood of any wild animal. The mother spends nine years raising each of her children.

Parental Care

An opposite reproductive strategy is to have only a few young at one time and invest time and resources into protecting them. This maximizes the chances of them reaching maturity and having a family of their own. Humans use this strategy. The babies of animals that use this system are often born very helpless and require the mother and father to carry them and find their food. Eventually, the offspring will learn to do this for itself.

All the baby aphids are genetically identical clones of their mother. They are already growing daughters of their own inside them as they are born.

Asexual reproduction means that aphids (also called greenfly) can spread fast, covering a plant in just a few days.

Digestion and Excretion

The body needs a frequent supply of chemicals for fuel and to use as raw ingredients for growing and maintaining the body. These materials come from our food, and digestion is the process that breaks down the food into useful substances. Waste materials are then removed using a process called excretion.

Digestive Tract

Digestion's job is to turn the complex substances in food into simpler chemicals that can be absorbed by the body. This process has many steps that take place in the digestive tract, which is a long tube that passes through the body from the mouth to anus. Chemicals called enzymes break up the food, and it is absorbed into the blood in the small intestine.

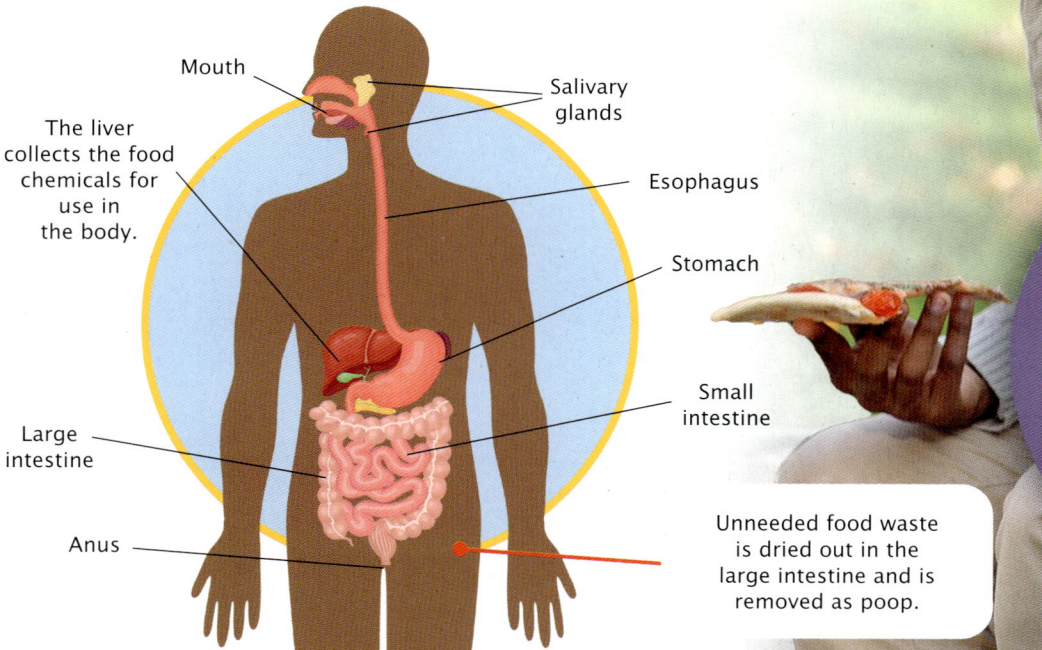

The liver collects the food chemicals for use in the body.

Mouth
Salivary glands
Esophagus
Stomach
Small intestine
Large intestine
Anus

Unneeded food waste is dried out in the large intestine and is removed as poop.

After swallowing, food is mashed up in the stomach. It is in there for about four hours as it is mixed with powerful chemicals that turn solid foods into a thick mushy liquid.

DID YOU KNOW? The small intestine is only 3.5 cm (1.5 inches) wide and about 7 m (23 ft) long, but its inner lining is covered in tiny hairlike extensions. If these extensions were made flat, they would cover a tennis court.

All the blood in the body passes through the kidneys every few minutes. The kidney filters the blood to remove damaging materials.

Excretion

Life processes taking place inside the body create waste products that have to be removed. The kidneys collect unwanted chemicals to make a water mixture called urine. Urine travels along tubes to the bladder, where it is stored. When the bladder is about half full, it creates the urge to urinate (or pee), and the urine is emptied out.

Digestion starts as the teeth chew up food into smaller chunks. Saliva is added in the mouth to make smooth balls of food that are easy to swallow.

HALL OF FAME: Santorio 1561–1636

The link between food and digestion was proven by a long experiment carried out by this Italian scientist over 30 years. Santorio built a weighing chair so he could keep a record of his weight before and after every meal and every time he used the toilet. He also weighed his meals, his urine, and his poop. Santorio's data showed that the weight of the food did not equal the weight of the waste. Some of the material from the food was going into his body.

Respiratory System

The body needs a constant supply of oxygen from the air. It is the job of the respiratory system to provide it. The respiratory system includes the airways and the lungs. Air coming into the lungs exchanges some of its oxygen with carbon dioxide, a waste gas that is breathed out.

Air entering the lungs is 21 percent oxygen. The air that comes out is only 16 percent oxygen.

Breathing Cycle

On average, you inhale (breathe in) and exhale (breathe out) every four seconds. The process is controlled automatically, mostly by a large muscle called the diaphragm. To breathe in, the diaphragm flattens. It stretches the lungs downward. Air flows in from the nose and mouth to fill the extra space. To breath out, the diaphragm bends upward and squeezes the air out.

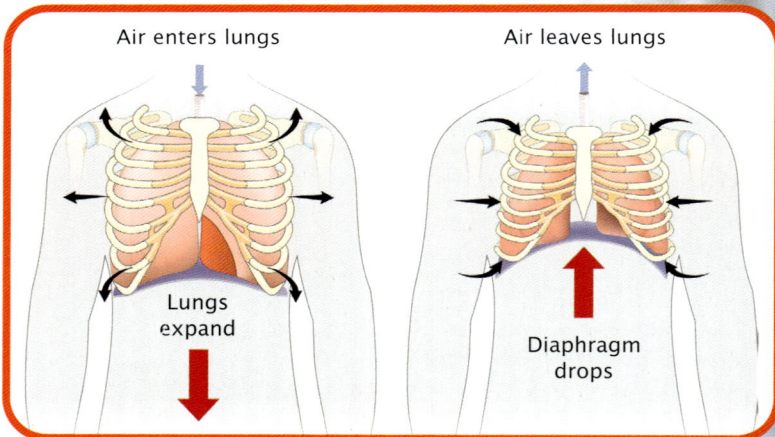

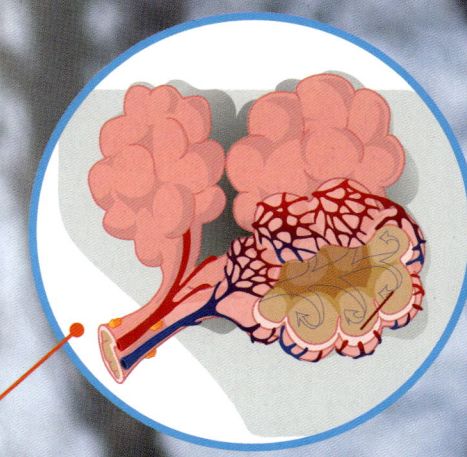

Each lung is full of sacs called alveoli that are surrounded by blood vessels. Oxygen in the air moves to the blood. Carbon dioxide in the blood moves the other way.

DID YOU KNOW? The world record for someone holding their breath belongs to Budimir Šobat from Croatia. In 2021, he breathed in pure oxygen and held his breath underwater for 24 minutes!

The breath also includes water vapor from the moist lining of the lungs. This vapor forms cloudy breath on cold days.

Coughs and Sneezes

When there is a blockage or an irritation of the airways, it is cleared with a cough or sneeze. These work by drawing in a big inbreathe and then closing off the airways using a flap in the throat called the epiglottis. (Normally the epiglottis's job is to keep food from going into the lungs.) With the epiglottis closed, the air pressure inside the lungs builds up. Once released, the surge of air will push the blockage out of the way.

During a sneeze, the tongue is used to block the mouth so that all the air is forced out of the nose.

On average, a human breathes 13,500 liters (2,970 gallons) of air every day. That's enough to fill 85 barrels.

HALL OF FAME: Galen
129–216

This Greek doctor had the job of treating wounded gladiators after deadly fights in the Colosseum in Rome. Galen got to see a lot of the internal structures of the human body. He also cut up dead animals, like pigs, to learn more. Galen used bellows to pump air into the lungs, and he was the first doctor to show how they were connected to the throat by a windpipe, or trachea. He also identified that the larynx, or voice box, is at top of the trachea.

Circulatory System

The blood supply is the human body's transportation system delivering oxygen and food to all body parts and taking away the waste. The heart pumps blood around the body along tubes or vessels. Together these body parts make up the circulatory system.

It is important to exercise regularly so that the heart and circulatory system stay healthy. Any exercise that makes you feel tired out will help.

Double Loop

The circulatory system is made of two loops of blood vessels that connect to the heart. The larger loop takes the blood around the body. The smaller loop connects the heart to the lungs, where the blood can collect fresh oxygen and discharge carbon dioxide.

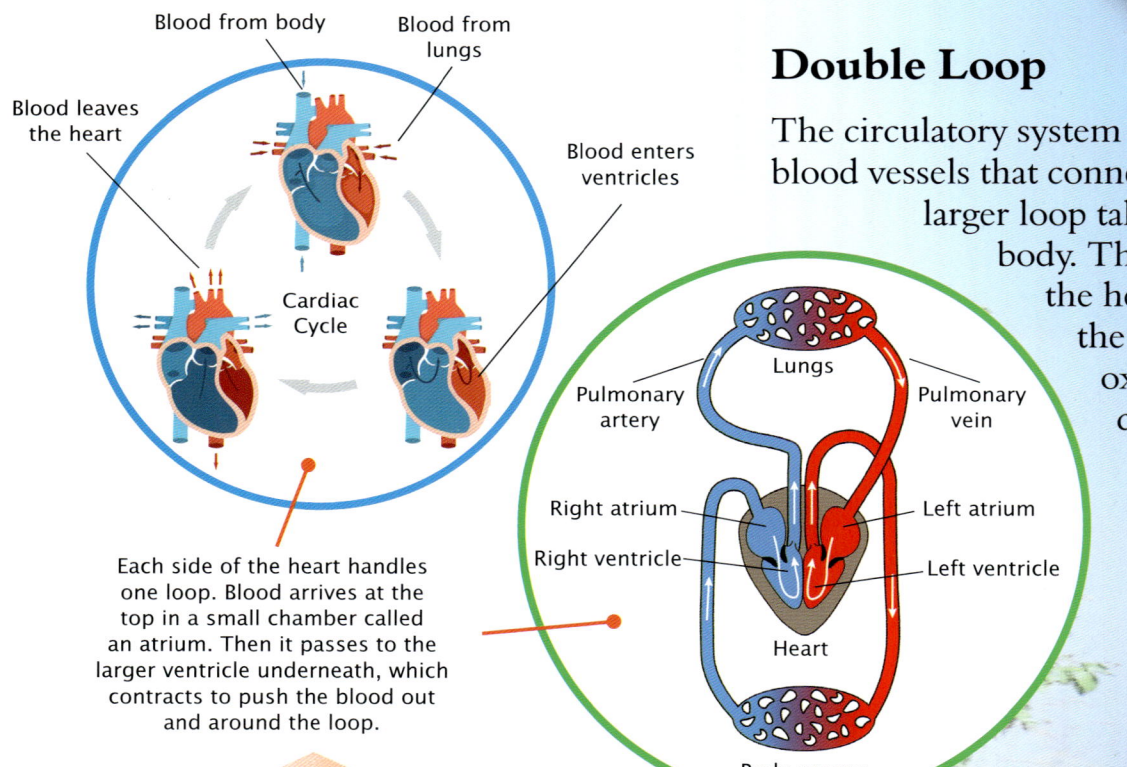

Each side of the heart handles one loop. Blood arrives at the top in a small chamber called an atrium. Then it passes to the larger ventricle underneath, which contracts to push the blood out and around the loop.

HALL OF FAME: William Harvey
1578–1657

Ancient doctors believed that the blood spread out from the heart and was absorbed by the body, and new blood is being made all the time. William Harvey, a doctor who looked after the British royal family, thought this did not make sense. He was not allowed to examine dead human bodies, so instead he experimented with animals. In 1628, he confirmed what others had already suspected—that blood circulates around the body in closed loops.

Skeleton

The human skeleton contains 206 bones. At birth, we have 270 bones, but as we grow older, several smaller bones fuse together to make larger ones. The skeleton is the internal framework of the body. It is there to give the body its shape, to create a protective cage around the soft organs, and to create solid anchor points for muscles and ligaments.

Joints

The place where two bones meet is called a joint. The human skeleton has 340 joints, although most of those are fixed and inflexible. All body movements happen as flexible or synovial joints. The bones are connected by elastic straps called ligaments. There are six kinds of synovial joints in the human body, each one able to move in different ways with twists, bends, and swivels.

The place where the bones meet is surrounded by a fluid-filled capsule. The tips of the bones are also padded with soft cartilage tissue.

Bone
Joint cavity filled with fluid
Cartilage
Bone
Muscle
Joint capsule
Tendon

The main body has the axial skeleton. This is made up of 80 bones including the flexible spinal column of 33 vertebrae, 24 ribs, and the skull, which is a brain case made of 21 bones fused together.

HALL OF FAME:
Mary Leakey
1913–1996

The Leakey family is famous for discovering fossil skeletons belonging to the distant ancestors of modern humans that lived in East Africa millions of years ago. Mary Leakey discovered the skeleton of an early relative of African apes, including chimpanzees, gorillas, and humans, that lived in the area about 20 million years ago.

DID YOU KNOW? The smallest bone in the body is not connected to the skeleton. It is the 2-mm (0.08-in)-long stapes, or stirrup bone, that is used to transmit sound through the ear.

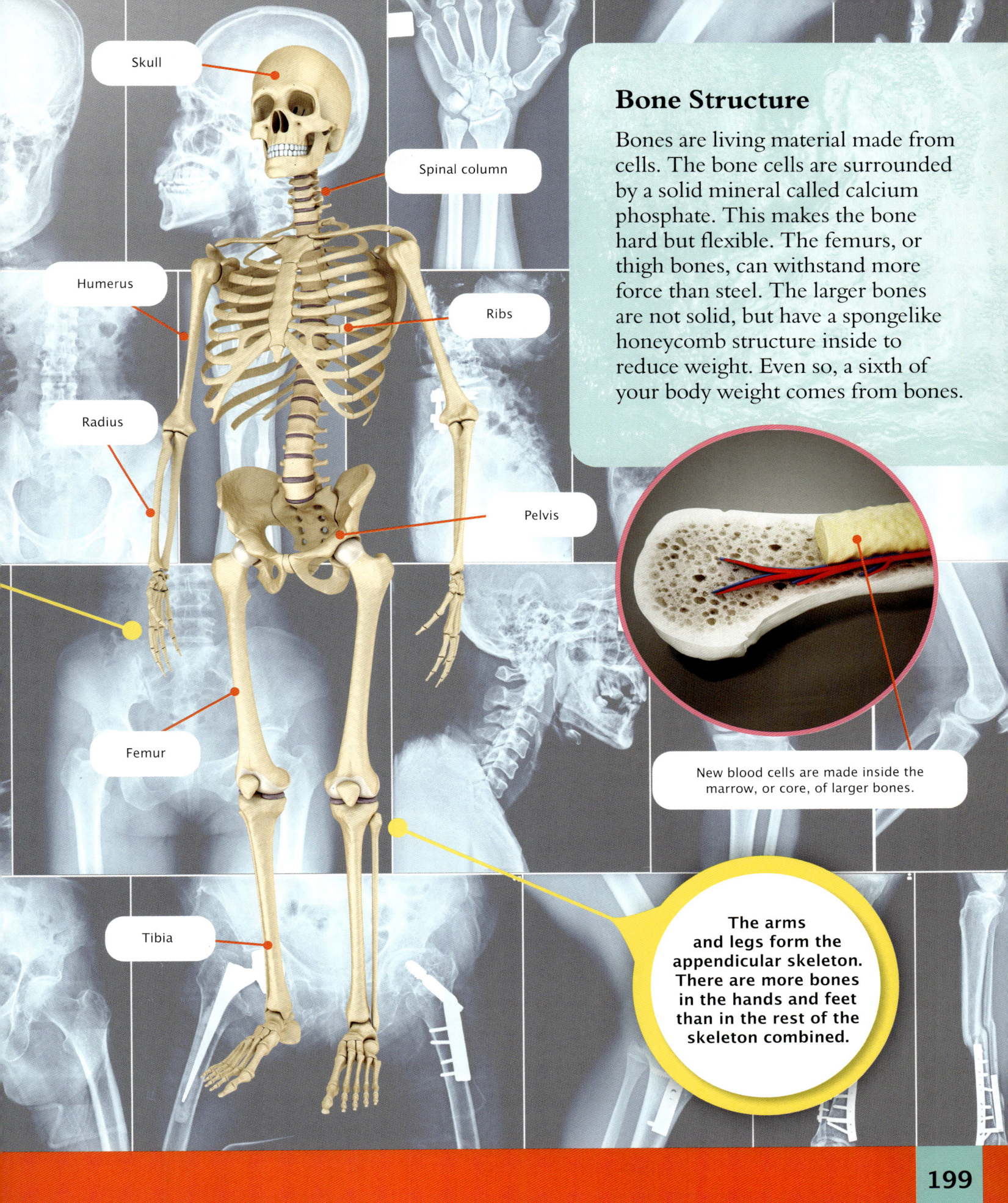

Bone Structure

Bones are living material made from cells. The bone cells are surrounded by a solid mineral called calcium phosphate. This makes the bone hard but flexible. The femurs, or thigh bones, can withstand more force than steel. The larger bones are not solid, but have a spongelike honeycomb structure inside to reduce weight. Even so, a sixth of your body weight comes from bones.

New blood cells are made inside the marrow, or core, of larger bones.

The arms and legs form the appendicular skeleton. There are more bones in the hands and feet than in the rest of the skeleton combined.

- Skull
- Spinal column
- Humerus
- Ribs
- Radius
- Pelvis
- Femur
- Tibia

Muscles

There are hundreds of muscles in the human body, consisting of three types. Cardiac muscle is only found in the heart, and it doesn't get tired like other muscles, so it can keep working all the time. Smooth muscle is used in the gut, arteries, and other tubes. Skeletal muscles, of which there are around 650, are used to move the body.

> Muscles cannot contract forever. The muscle cells create lactic acid as they work hard to contract, and this acid makes the muscle burn and feel tired. Eventually it must relax.

Moving Joints

Muscles create forces by contracting, or growing shorter, so they cannot push, only pull. As a result, skeletal muscles work in pairs to move joints, with one of the pair contracting while the other stays relaxed. One muscle, the flexor, contracts to bend the joint. On the other side of the joint, the extensor muscle contracts to straighten the joint.

> Muscles are made up of billions of microscopic protein fibers bundled together.

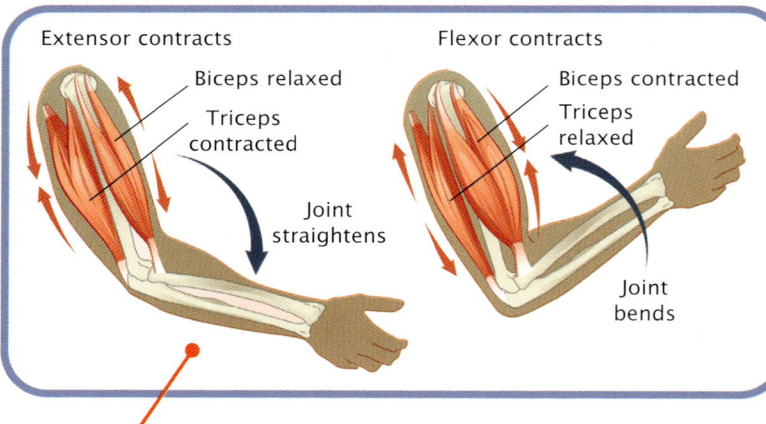

The muscle is attached to bones by connectors called tendons. The tendon does not stretch much at all, so all of the force from the muscle is transferred to moving the joint.

HALL OF FAME: Luigi Galvani
1737–1798

In 1780, Italian scientist Luigi Galvani found that the muscles of frogs contracted as an electric current flowed through them. Galvani thought that electricity was made by the living force of an animal—and it was still there, for a while at least, after it had died. The discovery led to the invention of the battery, and also much later it was shown how contractions were stimulated by the movements of electrical charges in muscle cells.

Smooth Muscle

The muscles that push food through the gut are called smooth muscle. It is less strong than skeletal muscles. Smooth muscle is either longitudinal, meaning its fibers all line up, or it is circular, so fibers are in rings. These two types work together to create rhythmic waves of contractions in the gut that push food along.

The waves of contractions in the gut are called peristalsis. For this to work best, the food in the gut needs to contain solid fiber, so the muscles have something to push against.

Exercising muscles makes them bigger and stronger. By working a muscle hard, the fibers become a little damaged. The muscle gets stiff as they mend, and they grow back bigger than before.

When a joint bends too much, the ligaments that connect the bones together become stretched. This injury is called a sprain. The best thing to do is rest the joint so the ligament can recover.

DID YOU KNOW? The smallest muscle in the body is the stapedius muscle—it is only 1 mm (0.04 in) long. It pulls on the tiny bones inside the ear, helping to reduce the effects of loud sounds.

Nervous System

The nervous system is the main way that different parts of the body communicate. It is in two parts. The central nervous system is made up of the brain and the spinal cord. The peripheral nervous system is a network of wirelike nerve cells that spreads through the body. These nerve cells collect information from the senses and send it to the central nervous system. The brain sends back a response through the nerves to the muscles and other body parts if necessary.

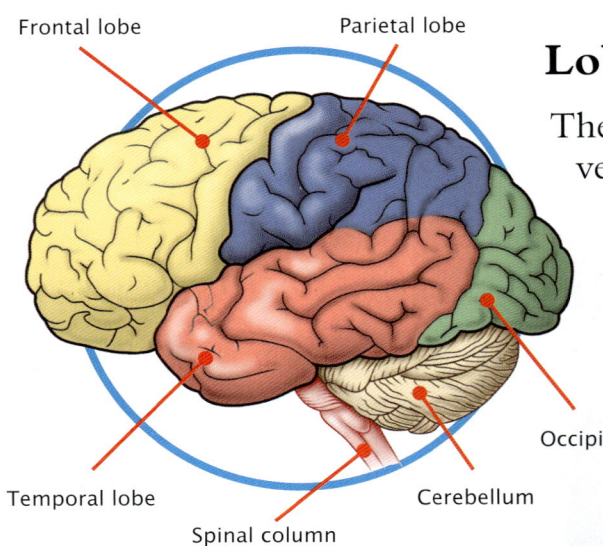

Frontal lobe · Parietal lobe · Occipital lobe · Cerebellum · Spinal column · Temporal lobe

Lobes of the Brain

The outer layer of the brain is called the cerebral cortex. It is very large in humans and is divided into four lobes, or regions. Thoughts and decision-making are handled by the frontal lobe. Vision is processed by the occipital lobe, while the parietal and temporal lobes are linked to speech, memory, and the other senses.

The cerebellum low down at the back of the brain is a filter system that ensures muscle movements are smooth and coordinated.

Reflex Action

Not all movements are controlled by the brain. Sometimes muscles work by a reflex, which is controlled by the spinal cord. The recoil reflex pulls the hand away from a sharp or hot object. A nerve signal does not go all the way to the brain but instead loops through the spinal cord to the arm muscle, creating a "reflex arc." The reflex ensures that action is taken as soon as possible to prevent injury.

Sensory nerves bring signals to the central nervous system, while motor nerves carry them away.

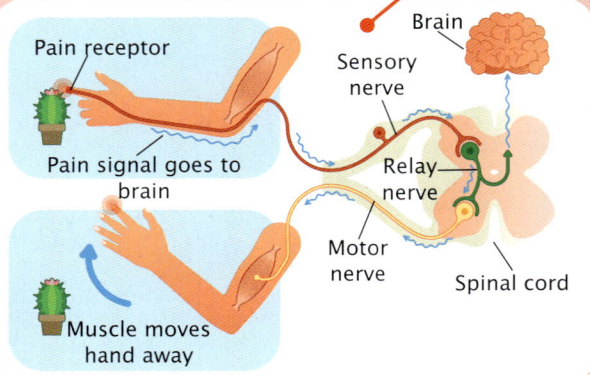

Pain receptor · Pain signal goes to brain · Muscle moves hand away · Brain · Sensory nerve · Relay nerve · Motor nerve · Spinal cord

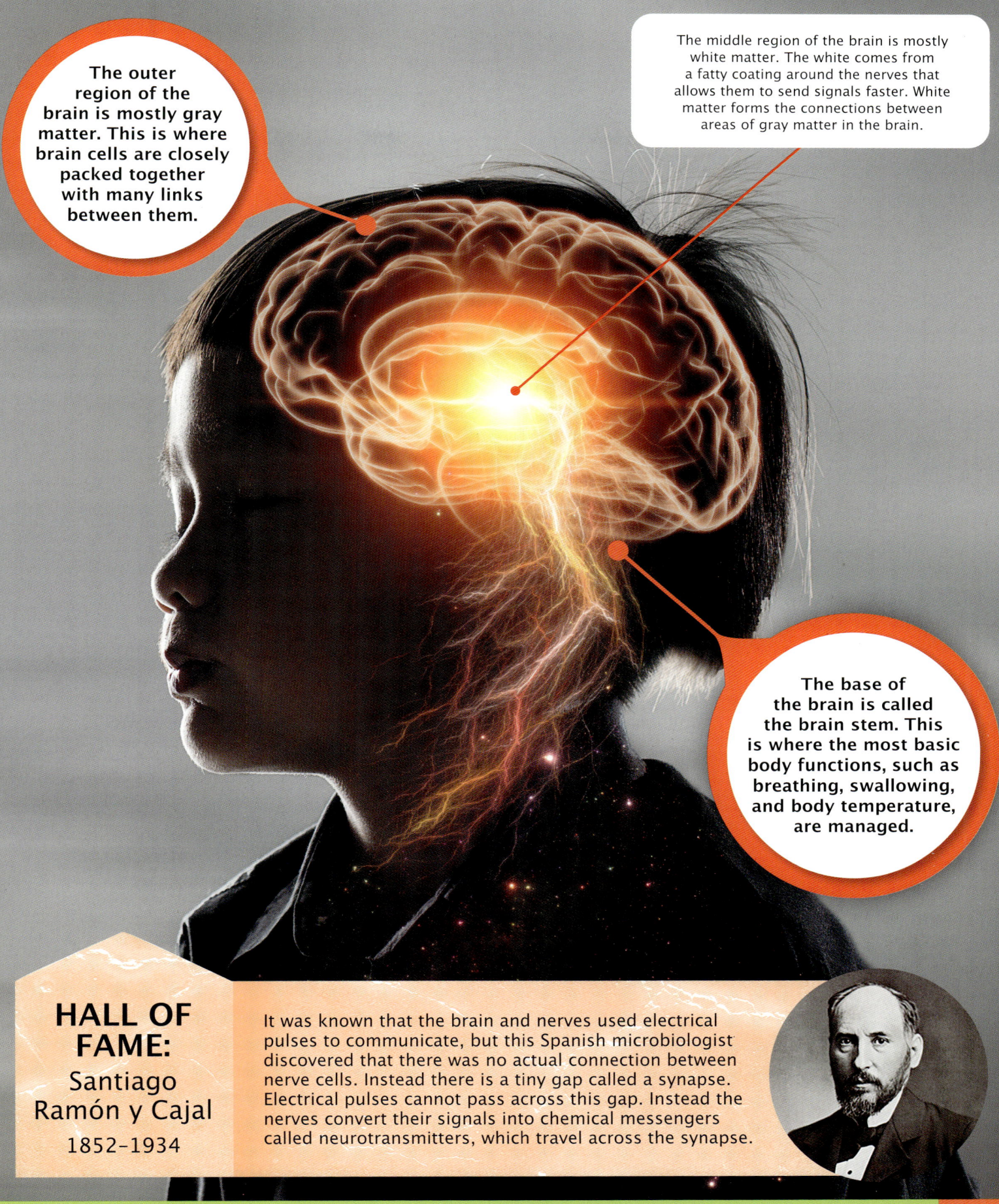

The outer region of the brain is mostly gray matter. This is where brain cells are closely packed together with many links between them.

The middle region of the brain is mostly white matter. The white comes from a fatty coating around the nerves that allows them to send signals faster. White matter forms the connections between areas of gray matter in the brain.

The base of the brain is called the brain stem. This is where the most basic body functions, such as breathing, swallowing, and body temperature, are managed.

HALL OF FAME: Santiago Ramón y Cajal 1852–1934

It was known that the brain and nerves used electrical pulses to communicate, but this Spanish microbiologist discovered that there was no actual connection between nerve cells. Instead there is a tiny gap called a synapse. Electrical pulses cannot pass across this gap. Instead the nerves convert their signals into chemical messengers called neurotransmitters, which travel across the synapse.

DID YOU KNOW? If the human brain was a computer, it would have a storage capacity of 2.5 petabytes, or 2.5 million gigabytes.

The Senses

The human body is said to have five senses: touch, sight, hearing, taste, and smell. However, this is simplifying the situation a lot. The human senses are constantly collecting information about what is going on inside and around the body. All this information is processed by the brain to create our sense of awareness.

There are millions of touch receptors in the skin that pick up different kinds of forces pushing on the skin, such as sharp pricks or hard pressure.

Sense of Vision

The eye works a lot like a camera to capture images. Light beams pass through the pupil at the front and are focused onto the light-sensitive retina at the back of the eye by the cornea and flexible lens. When light hits a cell in the retina, it stimulates a nerve signal. Many of these signals together create a record of the image formed by the eye, and this record is sent along the optic nerve to the brain for rocessing.

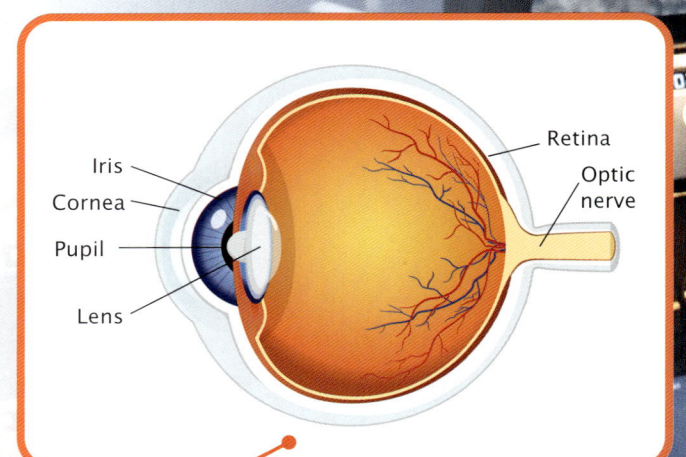

The iris can open and close to control the amount of light entering the eye. It opens wide in dark conditions and tightens up in bright light.

Sense of Hearing

Sound is the result of vibrations in the air. The ear is a highly sensitive touch organ that can pick up these waves, or movements in the air, and convert them into nerve signals. The wave enters the outer ear and makes the eardrum vibrate. That vibration is passed to three tiny bones, which tap out the vibration onto the cochlea. This is a spiral of fluid, and the sound wave ripples through it, wafting hairlike nerve cells that send out signals to the brain.

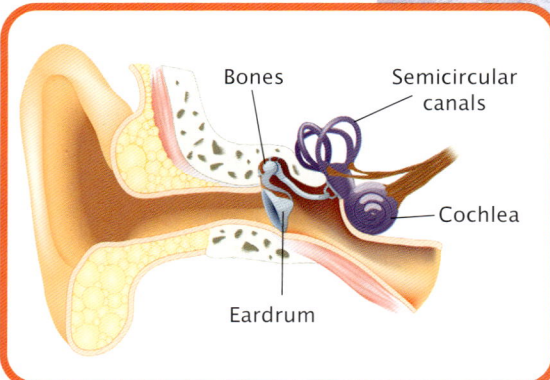

DID YOU KNOW? The human nose seldom detects one chemical at a time. Instead each odor is a mixture of many chemicals. The nose can distinguish around 1 trillion distinct odors.

Nasal cavity
Smell receptors
Nostril
Mouth
Tongue
Windpipe

The senses of smell and taste use chemical detectors on the tongue, gums, and inside the nasal cavity. The tongue can pick up at least five tastes, and the nose can pick up 10,000 distinct chemicals in the air.

The ear is also involved in the sense of balance. The brain detects changes in head position by the way fluid shifts the semicircular canals in the inner ear. When this fluid gets churned up, we feel dizzy.

HALL OF FAME: Ibn al-Haytham 965-1040

Also known as Alhazen, this medieval scientist from what is now Iraq conducted the first scientific experiments to show that the eyes detect beams of light arriving from the surroundings. Before Ibn al-Haytham's work, most people believed that the eyes sent out invisible flashes that scanned objects and reflected back. Al-Haytham's work with mirrors and lenses also helped to show how light was focused inside the eye into a sharp image of the scene in front of it.

Immune System and Diseases

The body is under constant attack from other organisms that are trying to get inside. Organisms that damage the body and cause diseases are called pathogens. They include bacteria, viruses, fungi, and even worms. The job of keeping pathogens out, and hunting down any that get in, belongs to the immune system.

The immune system needs to work fast to fight pathogens. It raises the body's temperature so the metabolism runs faster. A fever, or high temperature, is a clear sign that someone is sick.

Clotting

The skin is the first line of defense against attack. The outer layer is constantly shedding from the body and taking germs and dirt with it. Germs can get in through a cut in the skin, and this is sealed up quickly by clotting, where the blood forms a solid patch called a scab.

The scab is a network of solid protein strands. It gradually dries out and flakes off as the skin underneath is repaired.

White Blood Cell

If pathogens get into the body, it is the job of white blood cells to find and destroy them. There are several kinds of white blood cells. Some produce chemical markers called antibodies that stick to the attackers. Other blood cells then destroy anything with these marks. Memory cells keep a record of pathogens, so they can be dealt with if they infect the body again.

DID YOU KNOW? An allergy, such as hay fever or asthma, is caused by the immune system mistaking a harmless substances, like pollen, for a pathogen and responding to an attack.

The immune system uses a lot of the body's energy, which is why being sick makes us feel tired.

Pathogens can spread to all parts of the body. The lymph system is a set of tubes running through the body that drains liquids from the muscles and organs, and filters out pathogens.

HALL OF FAME: Ozlem Tureci
1967–

This German doctor is one of the scientists that created vaccines for Covid-19, a new disease that killed 7 million people between 2020 and 2023. Ozlem Tureci's vaccine was able to teach a person's immune system how to fight back against Covid, so they would be less sick and less likely to die from the disease.

Reproductive System

A human baby develops inside its mother's uterus, or womb, before being born. The baby starts life in a process called sexual intercourse. The penis, a male sex organ, enters the vagina, one of the female reproductive organs, and leaves behind sperm. This sperm moves to the uterus and combines with an egg cell produced by the mother. Together they make the first cell of a new human.

> Doctors who look after mother and child during pregnancy are called obstetricians. Midwives are also medical carers who are experts in helping people give birth.

Male Sex Organs

Sperm are produced inside the testes, small egg-shaped organs housed inside the scrotum, a sac hanging beneath the penis. The sperm are transported along tubes to the prostate gland, where they are combined with liquid called semen. During intercourse, the penis is filled with blood, which makes it longer and harder so it fits into the vagina, where the semen and sperm are released.

The penis also contains the urethra, a tube that connects to the bladder. Urine leaves the body through the urethra.

Female Sex Organs

Eggs, or ova, are made by rounded organs called the ovaries. Every month, an egg is released from one ovary, and it travels to the uterus via the fallopian tube. The uterus has been made ready for a baby to develop inside if this new egg meets a sperm. If not, the uterus will shed its lining. This process is called menstruation, or a period. It restarts the process of preparing for the next egg to arrive.

The opening of the uterus is called the cervix. During intercourse, the cervix allows sperm inside. It becomes tightly closed once a fetus starts to develop inside.

HALL OF FAME: Rebecca Lee Crumpler
1831–1895

In 1864, Rebecca Lee Crumpler became the first Black American woman to become a doctor. She became a specialist in child development and the care of women and babies after birth. She was working at a time after the American Civil War when enslaved people were being made free. Many white American doctors would not treat people who had been enslaved, and Crumpler worked for the government helping to provide them with care.

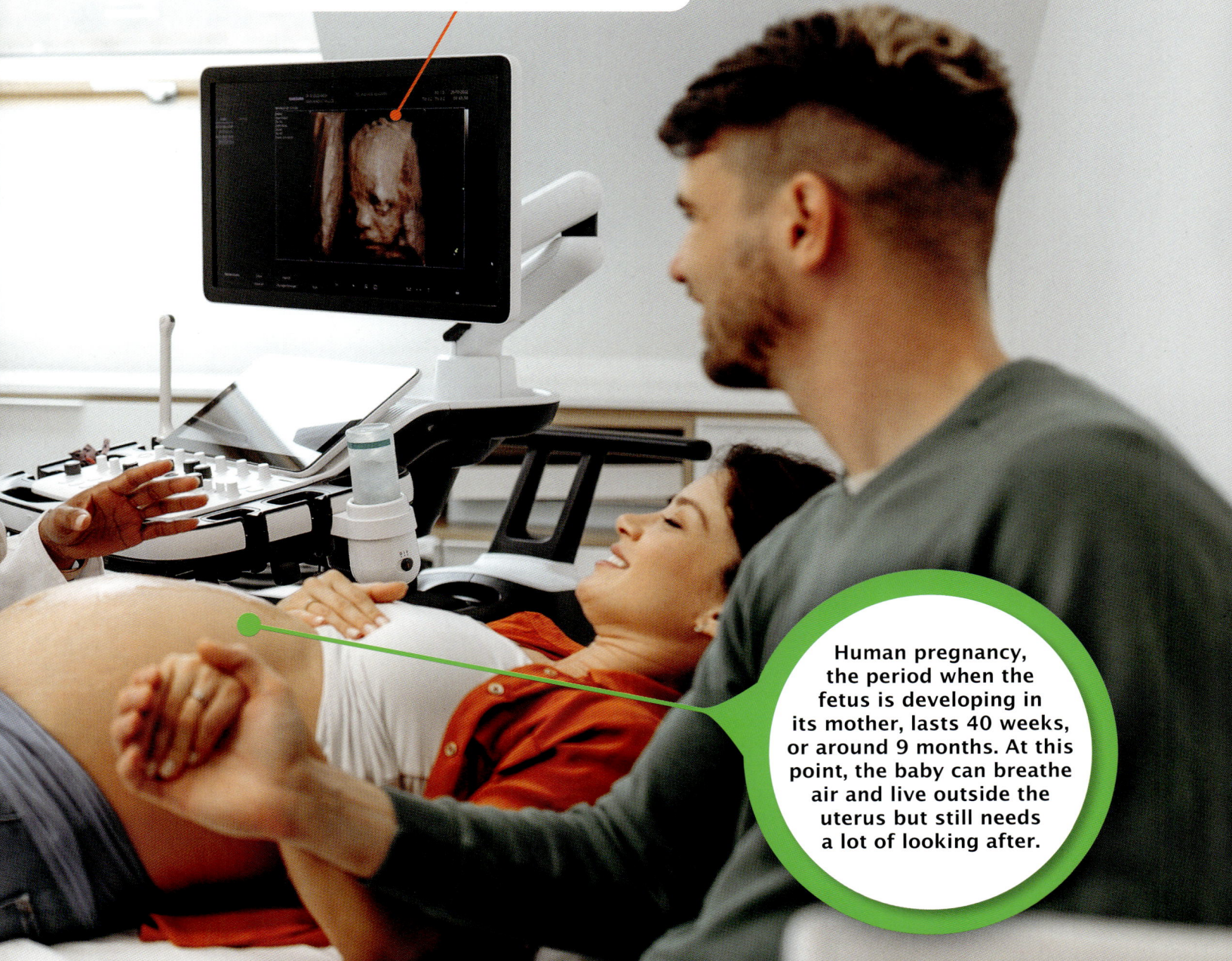

Ultrasound scanners send high-pitched sounds into the uterus. The sounds, which are harmless and too high to hear, echo off the fetus so the parents and medial experts can see it is healthy.

Human pregnancy, the period when the fetus is developing in its mother, lasts 40 weeks, or around 9 months. At this point, the baby can breathe air and live outside the uterus but still needs a lot of looking after.

DID YOU KNOW? In 2021, Halime Cissé from Mali became the only mother of nonuplets in history when she gave birth to five girls and four boys.

Chapter 10: Cells and Genetics

Studying Cells

One of the basic laws of biology is called cell theory. It says that every living thing has a body made of at least one cell—often many billions—and every one of those cells developed from an older cell. To understand how bodies work, we need to look more closely at cells.

The scientists that study cells and other tiny forms of life are called microbiologists. To see even more detail, they use electron microscopes.

Microscopes

The main tool for studying cells is the light microscope. It works by using two sets of lenses to magnify tiny objects so they can be seen in detail. Light shines up through the sample and is focused inside the microscope by the first lens into a tiny but highly detailed image. The lens in the eyepiece then magnifies that image so it is big enough for the human eye to see.

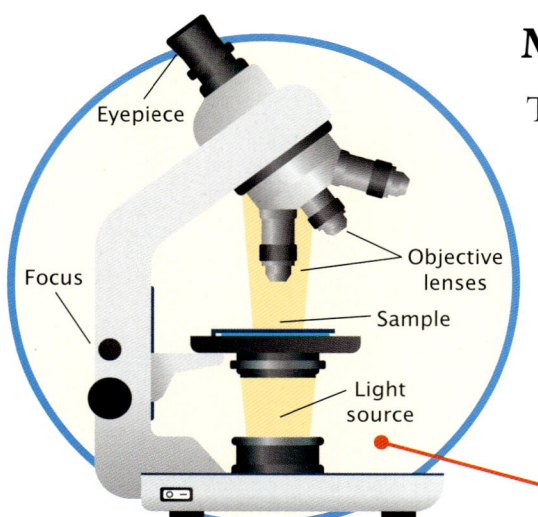

A biological microscope generally has three objective lenses, each giving a different level of magnification.

Preparing a Sample

The best way to examine cells is to place a very thin slice of tissue on a clear glass slide. This slice is bathed in a droplet of water, and a see-through cover is placed on top. This holds the sample still and flat so that the lenses can focus. Dyes are added to the water to highlight features of the sample. Salts and other chemicals can also be used to investigate how the cell operates.

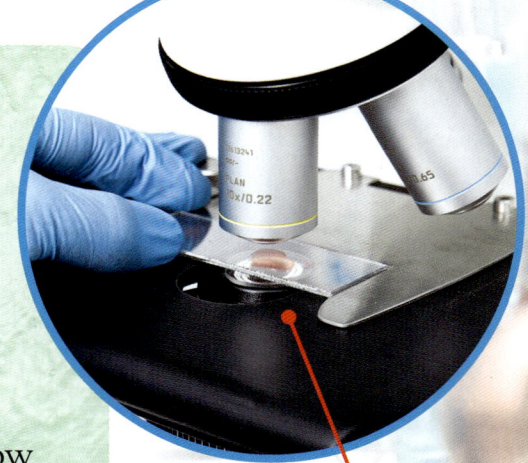

The sample is thin enough for light to shine right through. So the view through the microscope is a silhouette of the cells.

HALL OF FAME: Margaret Pittman 1901–1995

Margaret Pittman's start in science came as a child when she assisted her father, who was a doctor, with his patients. She excelled at school and college and worked as a teacher to save money to go to the University of Chicago, where she became an expert in bacteria and microbiology. She worked well into her 70s, investigating the bacteria involved in deadly diseases like cholera and meningitis.

The scientists can adjust the brightness and position of the light source to get a different view of the cell samples.

A pipette is used to add dyes or other chemicals to the sample.

DID YOU KNOW? A light microscope can see objects that are 200 times smaller than the width of a human hair.

Plant Cell

Plants and plantlike protists have a distinctive cell structure that sets them apart from other kingdoms of life. The most obvious feature is the cell wall that surrounds the cell's outer membrane and provides structural support. Additionally, the cells in the green parts of a plant have chloroplasts that are used for photosynthesis.

Internal Structures

All cells share some basic features. The contents of cells are in a watery liquid called cytoplasm, which is all contained inside an enveloping cell membrane. The cells of complex life, such as plants, have a nucleus where the DNA is stored. As well as chloroplasts and a cell wall, plant cells typically have a large vacuole. This is a bag used to store water, salts, and sugars.

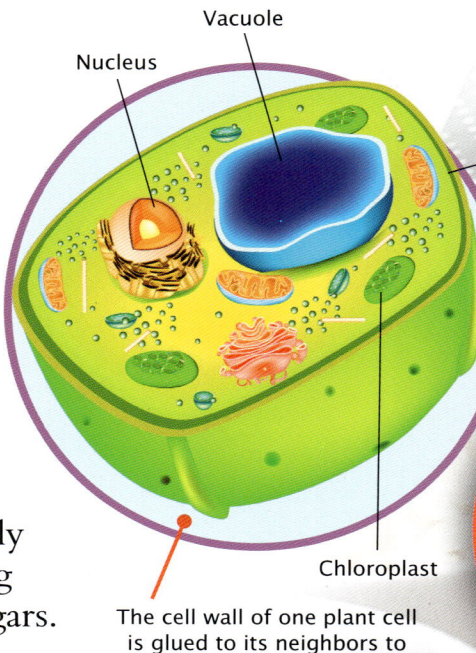

The cell wall of one plant cell is glued to its neighbors to create strong plant bodies.

The wooden board is composed of dead plant cells made of cellulose walls filled with another hard polymer called lignin.

Hay, or dried grass stalks, is almost pure cellulose. Farm animals have stomach bacteria that can digest this tough material.

Tough Cell Wall

The plant cell wall is made from a carbohydrate called cellulose. This is a polymer, or chainlike chemical, that is made from smaller glucose molecules linked together. The structure of cellulose makes it a very sturdy substance that makes plant bodies strong enough to stand upright. Cellulose is left behind after all the other parts of the cells have rotted away.

DID YOU KNOW? Plant cells are roughly rectangular. They are generally between 0.01 and 0.1 mm (0.0004–0.004 inches) long.

Animal Cell

An animal cell has no cell wall, just a flexible outer cell membrane. As a result, the cells have no standard shape and take many forms. Like the cells of a plant and fungus, animal cells have several types of internal structures called organelles.

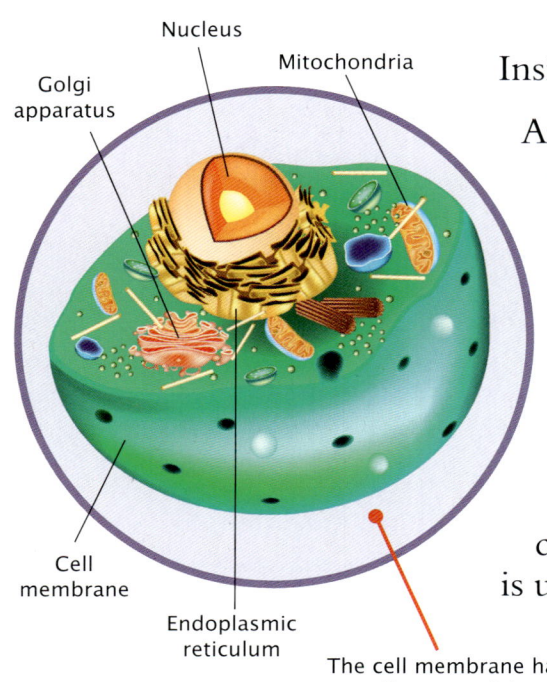

Inside the Cell

An animal cell contains several kinds of organelles, suspended within cytoplasm. As well as the nucleus, which stores the cell's genetic information and determines what the cell does, there are also the mitochondria for producing energy. Folded membranes called the endoplasmic reticulum are where the cell makes its useful chemicals, and the Golgi apparatus is used in transporting materials.

The cell membrane has pores and pumps in it that allow materials to enter and leave the cell.

The skin is also called the epidermis. Epidermal cells are dead and dried out and can form calluses. Trumpet players often get calluses on their lips.

Specialization

Animal bodies are made up of many cells that work together. The body cells are not all the same but are specialized to perform a certain job. Sponges are one of the simplest types of animals, with just four cell types involved in essential processes such as feeding and reproduction. There are more than 200 different types of cells in a human body. The specialized cells contain the basic set of organelles but may develop other features like flagella or cilia.

Sponges are filter feeders that draw water though their funnel-shaped bodies.

DID YOU KNOW? A bird's egg is a single cell. The ostrich egg, at 13 cm (5 inches) long, is the biggest animal cell in the world.

HALL OF FAME: Camillo Golgi 1843–1926

Italian microbiologist Camillo Golgi made many discoveries about animal cells, especially the shape and structure of nerve cells. The Golgi apparatus, an organelle in every complex cell, is named after him. He discovered faint traces of this organelle in the 1890s, but it was not fully described until the 1950s.

The cell nucleus shows up well, but smaller organelles are harder to see without a more powerful electron microscope.

The soft lining of the cheeks is covered in a loose layer of cells. These can be collected easily and viewed under the microscope.

Bacterial Cell

Bacteria have a much smaller and simpler cell than more complex organisms such as animals and plants. The most obvious difference is that there is no nucleus. Instead, the cell's DNA is floating in a rough bundle. Along with those of archaea, the bacterial cell is termed prokaryotic.

Types of Bacteria

There are two main shapes for bacterial cells. A rounded cell is called a coccus, while a rod-shaped bacterium is called a bacillus. When the bacteria form into long chains, they are called streptococci or streptobacilli. Cocci bacteria also form in clusters that are known as staphylococci. A pair of joined round bacteria are called diplococci. Less common cell shapes are bean shapes, comma shapes, long and thin filaments, and spirals.

There are about 30 trillion cells in the human body and about the same number of bacterial cells living on your skin and in your stomach.

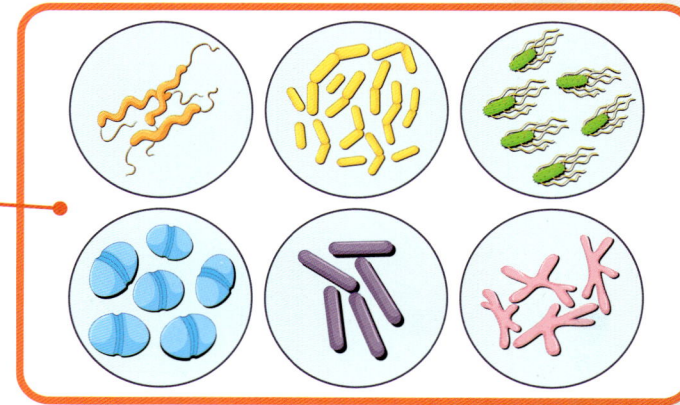

The names used to describe bacteria depend a lot on how the cells appear under the microscope. Another important way to identify different bacteria types is to use dyes that target certain chemicals in the cell.

HALL OF FAME: Alice Catherine Evans
1881–1975

Alice Catherine Evans was an American expert in bacteria. She worked for the United States government, studying disease spread in milk and cheese. Her work helped make these foods much safer. Evans also discovered which bacteria caused dangerous infections of the skin and blood in the 1940s, around the same time the first antibiotic medicines were being developed.

DID YOU KNOW? There are bacteria that eat rock. Some are found living 3 km (1.8 miles) underground and use the chemicals in the rock to provide energy.

Cell Features

All the life processes of the bacteria happen in the cytoplasm, which is a complex mixture of chemicals. The only obvious internal feature is the bundle of DNA. A membrane surrounds the cell, and it may include long taillike flagella or shorter extensions called pili. A cell wall surrounds the membrane, and in some cases the whole cell is inside a protective capsule.

The cell wall of a bacterium is made mostly from a complex sugar-based chemical called murein.

Yogurt is a food full of healthy bacteria. Bacteria can cause diseases, but they also help our digestion in many ways, breaking down foods that our stomach chemicals cannot.

Cell Membranes and Transport

Every cell is surrounded by a thin outer layer, or membrane. The membrane is made from fatty chemicals that form a barrier for large molecules, while smaller ones, like water and oxygen, can pass through. Cells rely largely on a physical process called diffusion, where substances naturally spread out from where they are common to where they are rare. However, some cells also use more active systems to move materials around.

Water is pulled into plants by osmosis. If there is not enough water in the plant cells, the cells become soft, and the plant body will wilt.

In and Out of Cells

A cell can release large quantities of a substance using a process called exocytosis. The substance is discharged by the Golgi apparatus into vesicles, or small membrane bags. The vesicle merges with the cell membrane, and the contents are outside the cell. Cells that secrete hormones or enzymes use exocytosis. Endocytosis is the process run in reverse, where material outside the cell is captured in a hollow section of the cell membrane, which then breaks off to form a vesicle inside the cell.

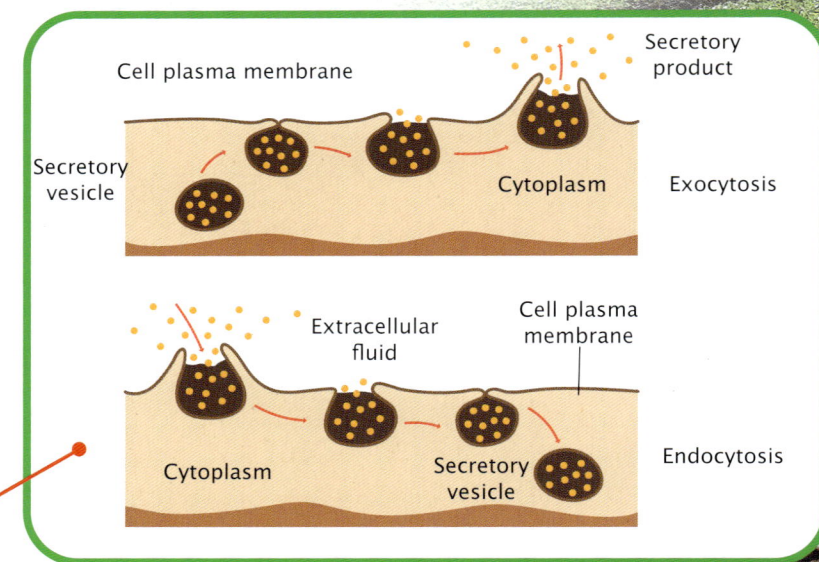

Endocytosis is used by cells that consume nutrients floating in the surrounding waters or that eat smaller cells.

DID YOU KNOW? Goblet cells secrete slimy mucus to coat the inside of the nose, lungs, and throat. In an adult human, they produce 1.5 liters (0.3 gallons) every day!

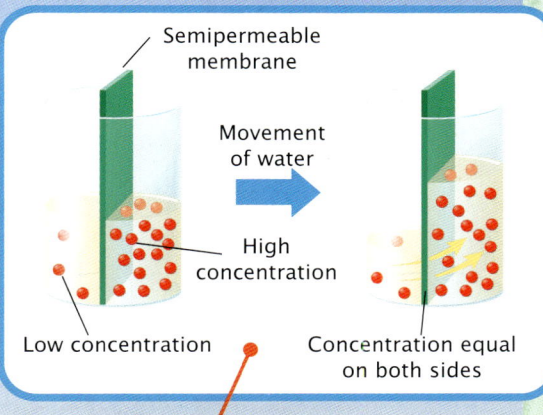

Water moves to make the concentration equal on both sides of the membrane.

Osmosis

Cells rely on a special kind of diffusion called osmosis to move water in and out of the cell. Water can cross the membrane, but other chemicals mixed into it cannot. As a result, when there is a high concentration of chemicals dissolved in the cytoplasm, water will diffuse in from outside to dilute it. In the same way, if the cell is in water that is more concentrated than the cytoplasm, osmosis will push water out of the cell, making it dry out.

Water is the universal solvent. This means that it is inside every cell, and all the chemicals needed for life are mixed into it.

HALL OF FAME:
Jean-Antoine Nollet
1700–1770

It would be impossible to understand how cells and living things worked without knowing about osmosis. It was discovered in 1748 by Jean-Antoine Nollet, a French priest, who put pure alcohol in a sealed pig's bladder that had been immersed in water. Several hours later, the bladder was bulging with water under great pressure. Osmosis had pushed water inside to dilute the alcohol.

Enzymes

The chemical reactions that take place in a living cell are called metabolism, and most of these reactions are regulated by enzymes. An enzyme is a biological catalyst, which means it is a chemical that allows a reaction to take place that would not happen by itself. There are many thousands of enzymes working in human cells.

Complex Shape

All enzymes are proteins, which are highly complex polymers, and each one has a uniquely intricate shape. Every enzyme has a specific job to do in metabolism, and that job is defined by the shape of the molecule. Its shape allows the enzyme to connect with other chemicals so they can react in some way. This idea is called the lock and key theory.

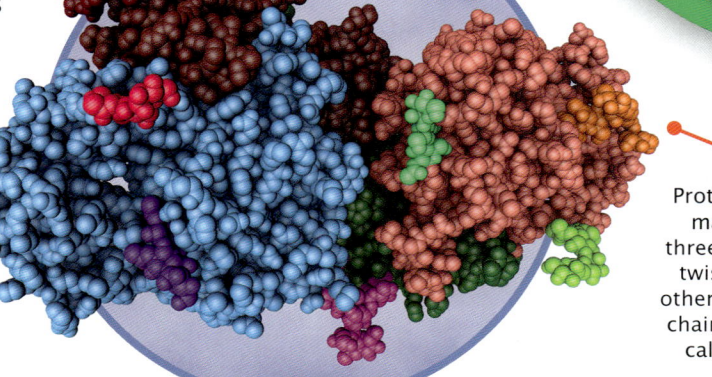

Fermented foods like pickled vegetables are made by enzymes released by yeasts or bacteria. The enzymes convert the sugars in the foods into acids, such as lactic acid or vinegars.

Protein molecules are made from two or three smaller polymers twisted around each other. The polymers are chains of smaller units called amino acids.

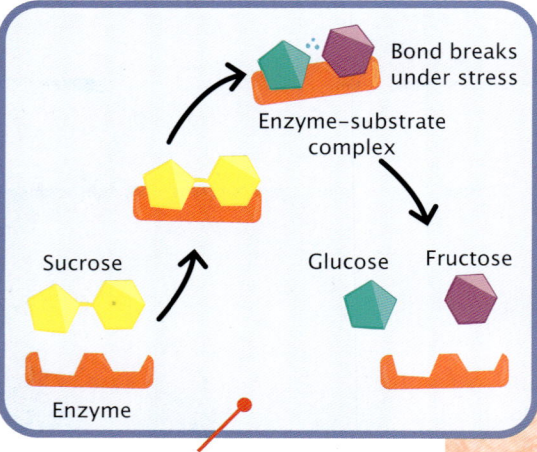

This enzyme is splitting a sucrose molecule into simple sugars. The enzyme is not used up in the process.

Lock and Key

In the lock and key theory, an enzyme has a region called the active site that is shaped exactly right for other molecules (called the substrate) to fit into—like a key in a lock. Once connected to the enzymes, the bonds inside the substrate molecules change in strength, so the atoms can rearrange in some way to create a new set of molecules called the products. The products are released from the enzyme, which is then ready to start again.

HALL OF FAME: AlphaFold 2021

The shape of an enzyme, or any other protein, is determined by hundreds of smaller units called amino acids. These are chained together in a very particular order and will push and pull on each other to fold up into the final protein shape. No human could figure out how to predict the shape of a protein from the order of its amino acids. However, in 2021, an artificial intelligence made by Google called AlphaFold was able to figure this out. Thanks to AlphaFold, it is now easier to read genes, and artificial enzymes can be designed for use as medicines.

Fermented foods taste sour and acidic due to the action of enzymes.

The large amount of acid in the fermented foods keeps other microscopic organisms from growing on the foods, so they do not rot or go bad very quickly.

DID YOU KNOW? There are over one million chemical reactions happening inside each of your cells every single second, and nearly every one requires an enzyme.

221

Cell Division

Cells can grow larger but will soon reach maximum size. For a body to grow, its cells need to divide in two—again and again. The cell division process used for growth like this is called mitosis. It transforms one parent cell into two almost identical offspring cells. Complex cells like those of plants and animals undergo mitosis. Bacteria use a similar system called binary fission.

> Between cell divisions, the cell is in interphase. During this time, the cell grows larger and organizes its chromosomes, ready for the next division.

Fast Growers

Cell division allows single-celled organisms to reproduce very quickly. For example, the microscopic algae that float in seawater as plankton can double in number every 24 hours. Soon there are so many that the invisible plantlike organisms have turned the water green. This explosion of life is called an algal bloom. It can cause problems by spreading poisons in the water and blocking light from reaching the water lower down.

Algal blooms are often caused by fertilizer chemicals from farms washing into water. The chemicals make the algae grow and divide much faster than normal.

Mitosis

There are several phases to a cell division by mitosis that ensure the offspring cells always have the same genes as the parent cell. The chromosomes in the nucleus are copied into double versions with an X shape. These are lined up in the middle of the cell, and then each equal half is pulled to opposite ends. Finally, a new cell membrane forms across the middle, and the cell splits into two.

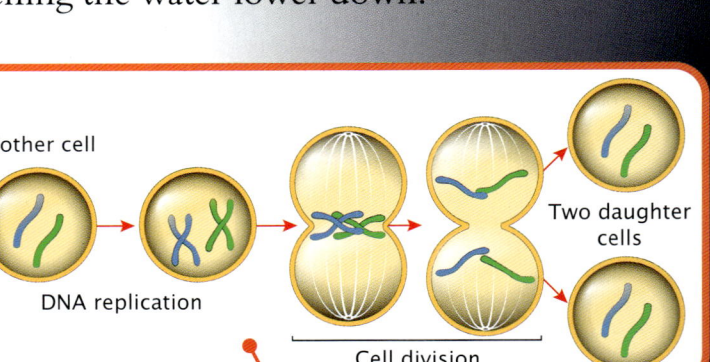

Mother cell · DNA replication · Cell division · Two daughter cells

> Each set of chromosomes is pulled to one end of the cell by microfilaments anchored there, separating the two sets. The organelles are divided equally between the two halves.

HALL OF FAME: Matthias Jakob Schleiden
1804–1881

Matthias Jacob Schleiden was one of the scientists behind cell theory. He had started work as a lawyer, but this made him unhappy—so he switched to studying the new field of cell biology. He was interested in cell division and showed that the contents of the nucleus were always shared by the new cells. He was also interested in evolution and was one of the first biologists to accept Charles Darwin's theory of evolution in 1859.

In the final stage of cell division, known as cytokinesis, a new membrane forms in the middle of the cell dividing up the cytoplasm.

Once the division is complete, a new nucleus forms around the chromosomes in the cell.

DID YOU KNOW? A bacterium can divide in two every 20 minutes. In just 7 hours, one bacteria can grow into more than 2 million.

DNA and Chromosomes

DNA, short for deoxyribonucleic acid, is a chemical stored inside the nucleus of a cell on structures called chromosomes. It carries the organism's genes, which are coded instructions on how to build a new cell and grow an entire body.

Chromosomes

DNA is quite a delicate substance. It is protected inside the cell's nucleus, where it is kept separate from other chemicals that might damage it and alter its genetic coding. The long strands are coiled up to make bundles called chromosomes. The DNA is only uncoiled when it is being copied and decoded. The chromosome is mostly coiled and compact during cell division. The rest of the time, it thins out into finer strands.

A human cell has 46 chromosomes, but that number varies a lot from species to species. Half of the chromosomes come from each of the parents.

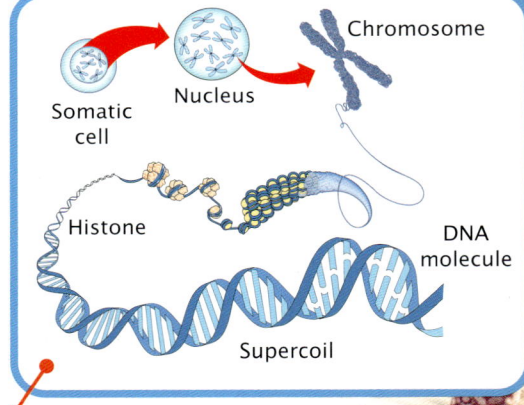

The DNA strands are coiled around support proteins called histones, and then these coils are coiled and coiled many times to make a compact "supercoil."

HALL OF FAME: Rosalind Franklin
1920–1958

Rosalind Franklin was a chemist who used X-rays to figure out the structure of molecules. DNA had been discovered in the 1860s, but 90 years later no one knew its shape. Franklin's X-ray photographs offered the first clue that the molecule was a helix (a shape like a spiral ladder). This was the big breakthrough that allowed other scientists to figure out how DNA works.

DID YOU KNOW? If all the DNA in your body was uncoiled, it would stretch from the Earth to the Sun and back 20 times!

DNA Structure

Deoxyribonucleic acid is a polymer built from several units to make a twisted ladder shape or helix. Ribose sugar forms the sides. There are four nucleic acids that connect in pairs to form the "rungs." The order of these acids along the DNA strand spells out a four-letter code. Genes are written in this code.

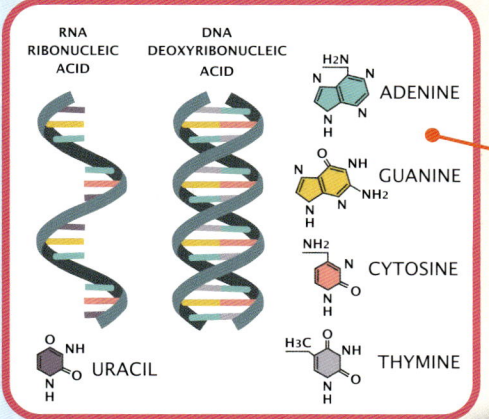

The four nucleic acids in DNA are simplified to the letters A, G, C, and T. RNA, or ribonucleic acid, is constructed from a single nucleic acid strand and uses the acid U instead of T.

The parental genes are altered and shuffled into new combinations for the offspring. This is how children inherit the features of their parents.

A full set of chromosomes is called a karyotype. Each chromosome is part of a pair, with one each coming from the mother and father. Humans have 23 pairs.

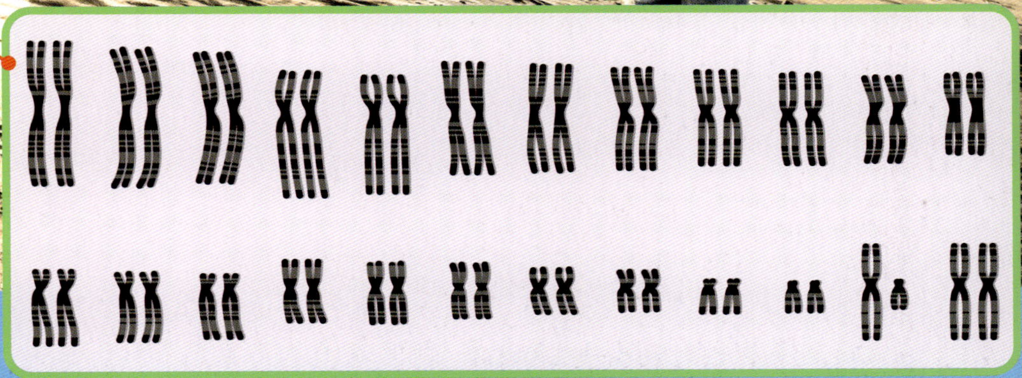

Meiosis

A special kind of cell division, called meiosis, is needed for sexual reproduction. Meiosis makes cells with a half set of genes—these are called sex cells. Two sex cells can merge to make a full set of genes for a new individual. This method of breeding is called sexual reproduction, and it ensures that each child has a highly varied set of genes.

> The four offspring cells created by meiosis can be used as sex cells, either sperm and eggs, in reproduction. Often only one of the four makes it to this stage.

Steps

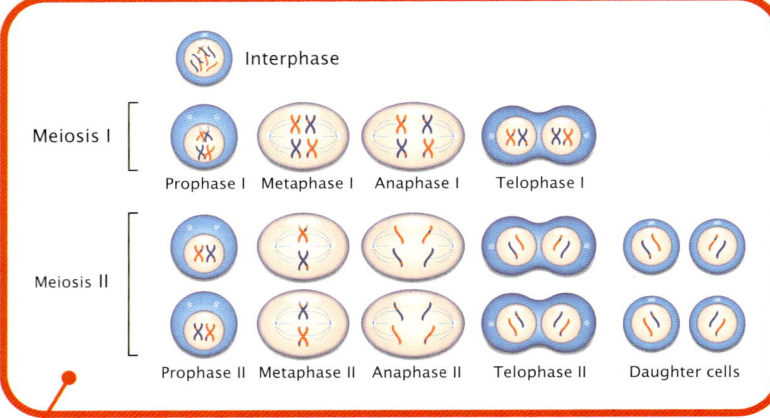

The starting cell is described as diploid, because it has a full set of paired up chromosomes inherited from two parents. The daughter cells made by meiosis are haploid, meaning they have just one of each of the pairs of chromosomes.

Meiosis is two cell divisions in one. The first division organizes the chromosomes into their pairs and then draws one of each pair to opposite ends of the cell. The cell then divides in two and creates daughter cells with a half set of chromosomes. The next division is more like mitosis (see page 222), and the two half-set cells divide again to make a total of four daughter cells.

Crossing Over

Unlike in mitosis, where the daughter cells always have identical genes, meiosis is deliberately mixing chromosomes up to make four daughter cells with a unique set of DNA. One of the ways this is done is called recombination, or crossing over. During the first division in meiosis, the paired up chromosomes are lined up next to each other. They are so close that they can tangle up and swap chunks of DNA.

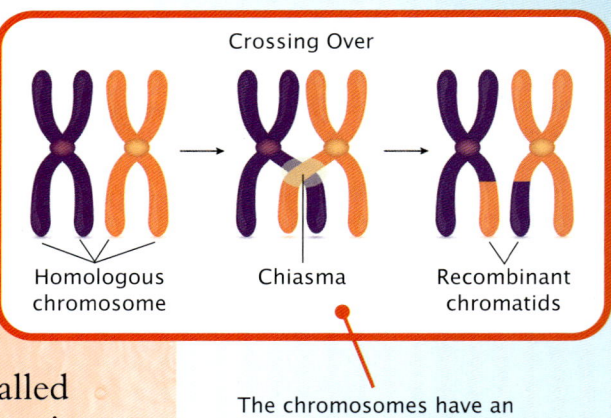

The chromosomes have an X-shape because they have been duplicated ready for cell division. They have two identical copies connected in the middle. During crossing over, only one of these halves swaps DNA.

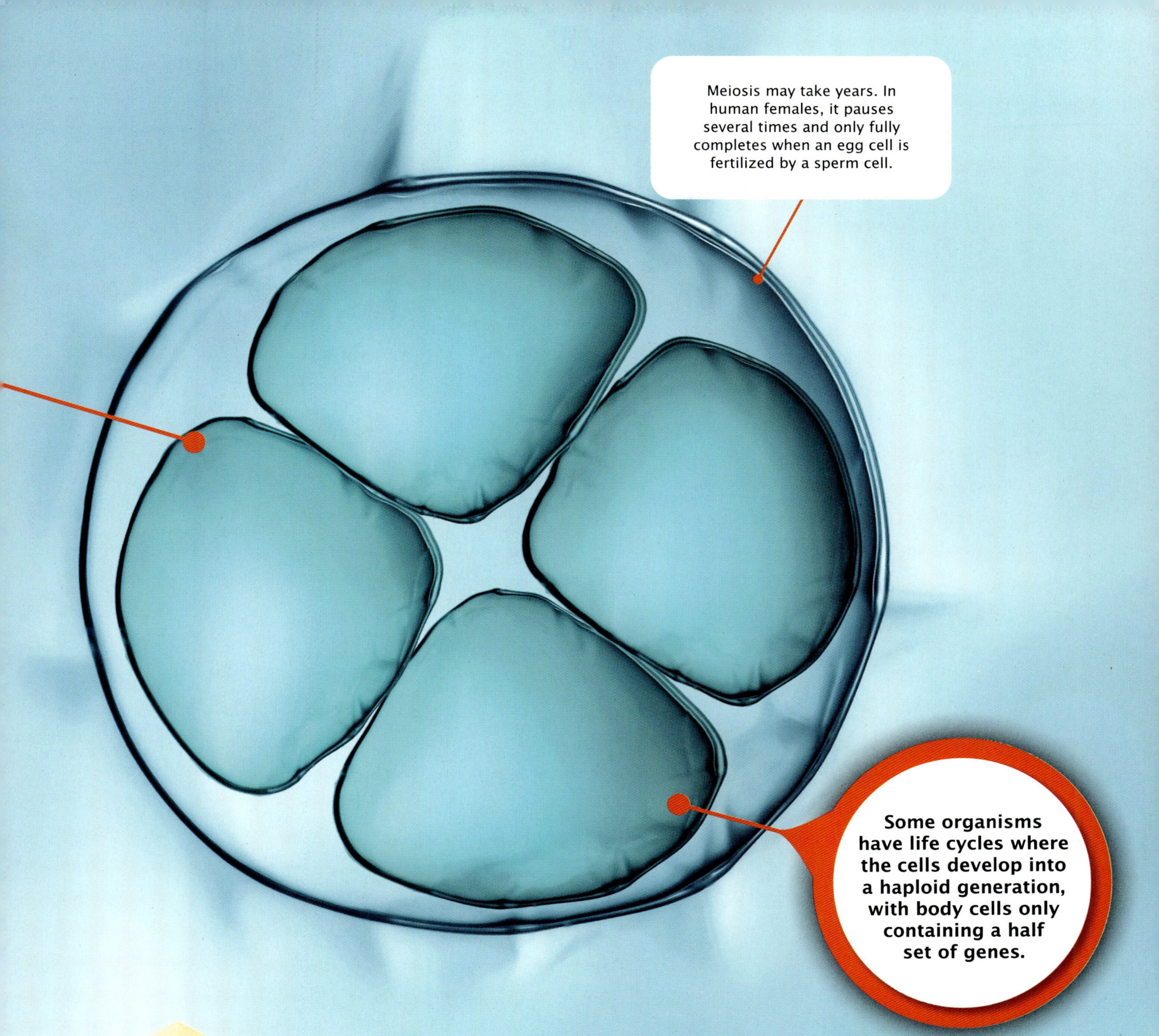

Meiosis may take years. In human females, it pauses several times and only fully completes when an egg cell is fertilized by a sperm cell.

Some organisms have life cycles where the cells develop into a haploid generation, with body cells only containing a half set of genes.

HALL OF FAME: Barbara McClintock
1902–1992

Starting out as a botanist studying crops grown on American farms, Barbara McClintock discovered recombination, one of the most important phenomena in genetics. She made this breakthrough while studying the chromosomes of maize, observing that they were muddled up during meiosis. Later, McClintock also made breakthroughs on how genes were read and expressed as phenotypes.

DID YOU KNOW? A female oceanic sunfish, or *Mola mola*, produces 300 million eggs each breeding season, which is the most for any vertebrate.

Sex Cells and Fertilization

Most multicellular organisms reproduce sexually. This involves two types of sex cells, or gametes, fusing together in a process called fertilization. The female gamete is the egg, and the male one is the sperm. Together they make a zygote, which is the first cell of a new individual.

> The human sperm is only 0.005 mm (0.0002 in) wide. The human egg is 20 times bigger, and at 0.1 mm (0.004 in) across, it is just about visible with the naked eye.

Gametes

Sperm and eggs are haploid, which means they only have half a set of chromosomes. The zygote they produce is diploid and has a full set. The sperm is built for swimming and has a long flagellum. It carries only energy supplies and DNA. The egg is much larger and cannot move itself. As well as some DNA, the egg contains everything else the new zygote will need to begin its life.

Egg

Sperm

HALL OF FAME: Oscar Hertwig 1849–1922

In 1876, German zoologist Oscar Hertwig discovered the process of fertilization by watching through a microscope as the sperm and egg of sea urchins fused together. Hertwig also discovered the process of meiosis. Additionally, he noticed that the chemicals in the nucleus were passed from cell to cell and must be the way characteristics are inherited. This was true, but it took almost another 100 years to figure out how!

DID YOU KNOW? The average female human releases about 350 ripe eggs in her whole life. An adult male human produces about 100 million sperm every day.

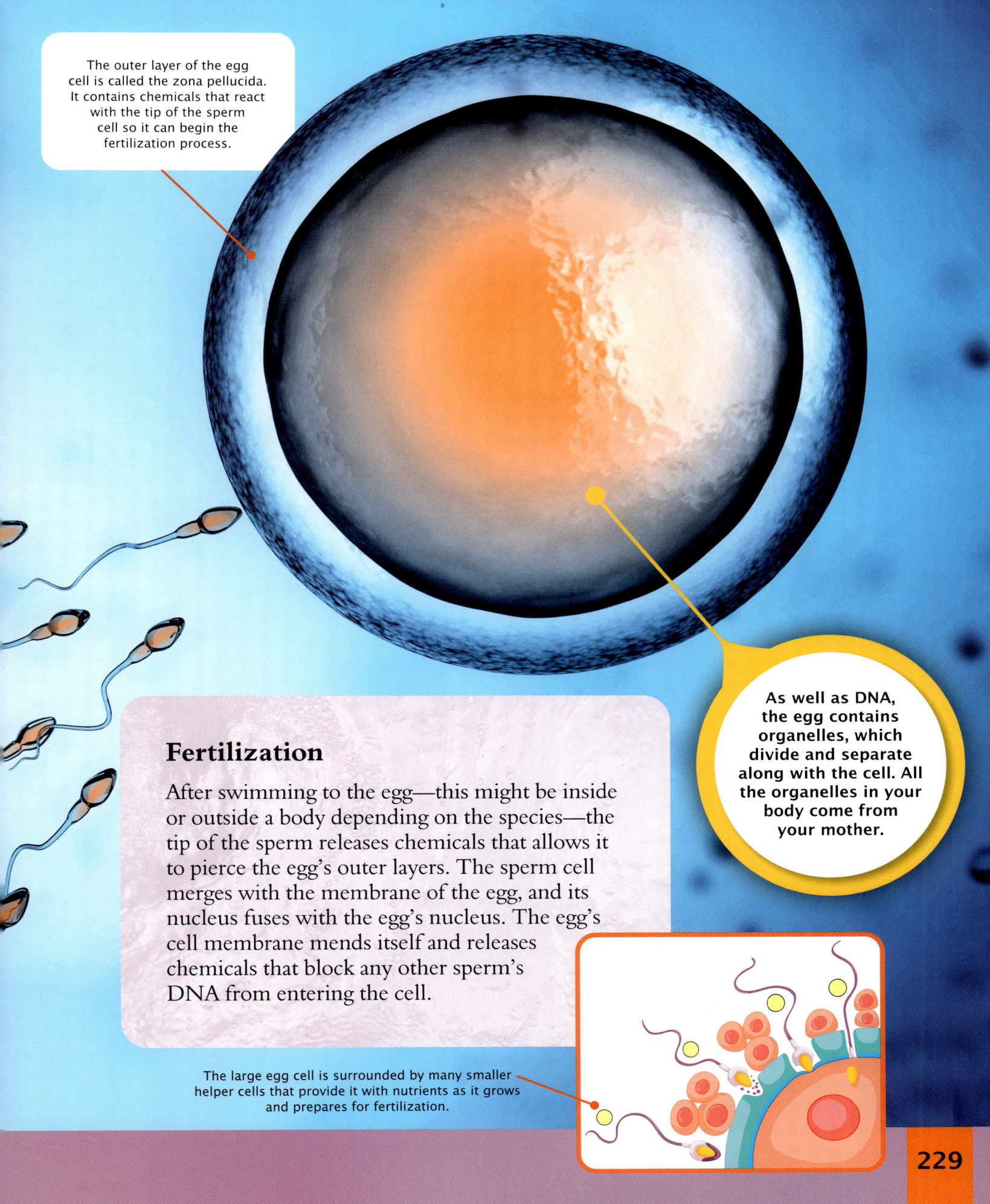

The outer layer of the egg cell is called the zona pellucida. It contains chemicals that react with the tip of the sperm cell so it can begin the fertilization process.

As well as DNA, the egg contains organelles, which divide and separate along with the cell. All the organelles in your body come from your mother.

Fertilization

After swimming to the egg—this might be inside or outside a body depending on the species—the tip of the sperm releases chemicals that allows it to pierce the egg's outer layers. The sperm cell merges with the membrane of the egg, and its nucleus fuses with the egg's nucleus. The egg's cell membrane mends itself and releases chemicals that block any other sperm's DNA from entering the cell.

The large egg cell is surrounded by many smaller helper cells that provide it with nutrients as it grows and prepares for fertilization.

The Theory of Evolution

The organisms that live on Earth today were not always here. Instead they evolved from earlier life-forms that have now become extinct and disappeared. Evolution is a system of change that is driven by a process called natural selection.

> This frog has failed to avoid being captured by a predator. Its genes and characteristics will not be passed on to the next generation.

Fossils

We know that different animals and plants lived long ago because of fossils. Fossils are the remains of living things that have turned to stone. They provide a record of how life has evolved slowly over many millions of years. The fossils also help to show how the environment was changing, and these changes are the major force that drives evolution. Simple life first appeared at least 3.5 billion years ago. All life that exists today—and all the extinct life seen in fossils—evolved from those early life-forms.

This plesiosaur was a reptile and relative of the dinosaurs. It lived in the oceans around 100 million years ago. It evolved from an older animal that lived on land.

HALL OF FAME: Charles Darwin
1809-1882

English naturalist Charles Darwin is world famous for publishing his book On the Origin of Species in 1859, in which he set out his theory of evolution by natural selection. At the time, many people were shocked by Darwin's ideas, but many years of research show that the theory is definitely the way life is able to change gradually to adapt to new habitats and conditions.

Natural Selection

All living things compete with each other for food and living space to survive. This struggle drives evolution using a process called natural selection. No group of organisms is identical; there is always variation. Some are a better fit for surviving in their habitat. These "fit" individuals do well and have many offspring, which are also "fit." Organisms that are less well suited for their habitat die without reproducing. This means that useful characteristics gradually spread through a population—and so species evolve slowly over time. The history of life is made of many tiny steps of evolution like this. Over billions of years, they have created the wealth of life on Earth.

The theory of evolution by natural selection was thought up by Charles Darwin as he traveled the world on HMS Beagle. On his voyage, he saw many unusual animals and plants.

The genes that this heron has inherited mean that it is a successful hunter—today at least. The better it is at hunting, the more likely it is to pass on those "fit" genes to the next generation.

Webbed feet are a useful characteristic for animals that swim. The frogs that live on land have evolved toes without webs.

DID YOU KNOW? Charles Darwin probably got many of his ideas about how life-forms were all related to each other from his grandfather Erasmus Darwin, who also wrote about evolution.

Sexual Selection

Natural selection is a process that allows only the organisms best suited to their environment to survive. How does this explain why some animals have features like bright feathers or huge tails, which make it harder for them to survive? The answer is a special kind of evolution called sexual selection.

> A male bird of paradise from New Guinea needs to spend a lot of time and energy getting noticed among the dense leaves.

Sexual Dimorphism

There are often obvious differences between the males and females of certain species. Features particular to one sex may be used to signal fitness to members of the other sex. Such features may include anatomical curiosities that have no obvious use for survival, such as colorful tails or huge antlers. The message here seems to be—if I can survive with something this impractical attached to me, I really must be fit.

This male duck is out to impress the female duck, showing off his plumage. She will choose which male bird to mate with this year.

Females that mate with big-headed males like this will produce male offspring that also have wide heads. In this way, the sexually selected feature spreads through the population.

Big Heads

This stalk-eyed fly has a very wide head with bulging eyes on each side. The males uses this feature to figure out which one is in charge and has the right to mate with a female. They line up head to head, and the male with the widest stalks wins. This system prevents risky fights. The loser will leave and look for another mate. Next time, he might be the winner in the same head-to-head competition.

DID YOU KNOW? Deer stags use antlers to signal who is the fittest. The Irish elk is an extinct deer species that had antlers 3.5 m (11.5 ft) wide, about the same length as a hatchback car.

The spiraled tail feathers or streamers catch the eye but also make it harder for the bird to fly. Only the fittest birds can stay strong and healthy despite these unhelpful features.

The bird relies on a courtship display to attract a mate. The females watch the males show off before opting for the best performing mate.

HALL OF FAME: Richard Dawkins 1941-

Richard Dawkins, a British professor of zoology, introduced the public to the ideas of neo-Darwinism, sometimes called the selfish gene. These ideas emerged in the 1960s and seek to explain how everything that happens in evolution is driven by the need for DNA to make copies of itself. Natural selection is really working at the genetic level, and living things are just survival machines built by DNA to help it copy itself.

Chapter 11: Habitats and Ecosystems

Ecosystems

Living organisms cannot survive alone. They require a place to live and a source of energy and nutrients. In all the places on Earth where life can survive, there is an ecosystem—a community of living things that rely on each other for survival. Evolution has given organisms the tools to survive in even the harshest environments.

A coral reef is a very diverse ecosystem in warm, shallow ocean waters, filled with shellfish, fish, and corals. Coral reefs are sometimes described as the rain forests of the sea due to the many species found there.

Ecological Factors

The study of ecosystems is called ecology. Ecologists have found that each ecosystem has a unique collection of factors that help or hinder the survival of its members. Biological factors come from other species living in the area. For example, one species may be a source of food, and others may pose problems, such as predation, disease, or competition for living space. The ecosystem also has nonbiological, or abiotic factors, such as weather changes and types of soils.

Ecosystems usually experience different seasons throughout the year, where changes in temperature, day length, and rain levels impact the wildlife.

HALL OF FAME:
Eduard Suess
1831–1914

Ecosystems of different kinds fill all of Earth's living space or biosphere. The term "biosphere" was coined in 1875 by Eduard Suess, an Austrian geologist. The biosphere is the part of the planet from the deep rocks to high atmosphere where life can survive. This made Suess one of the first ecologists. In the 1920s, his ideas were rediscovered by scientists who were trying to understand how wildlife communities worked.

Corals are tiny relatives of jellyfish. They have hard shells that are left behind when the soft parts die away. Younger corals grow on these shells, gradually building a region of rocky seabed filled with life.

Hibernation

In regions with long, cold winters, many animals will become inactive. They sleep a lot of the time and stay inside a warm den to save energy. There is not much food to be had outside, so the animals rely on a store of fat built up in the body through the warmer months. This is often called hibernation, although in true hibernation, an animal's body processes slow down, and its breathing and heart rate become very slow.

Reef corals are found in shallow, sunlit water because they have algae living inside them. The single-celled algae feed the corals with sugars made by photosynthesis.

Hedgehogs roll up in a ball of leaves to make a quiet, cozy nest for winter.

DID YOU KNOW? It is estimated that about 25 percent of all marine species are found in coral reefs.

Food Webs

All living things need energy and nutrients. Plants get theirs by harnessing the energy in sunlight to produce fuel using photosynthesis. Animals get what they need by eating, or consuming, the bodies of plants and other animals. A food web is a set of connections between members of an ecosystem, based on what is eating what.

Interconnected

A food web begins with a producer, which captures a source of energy from the wider environment. In almost all ecosystems, the producers are plants. The producers are eaten by primary consumers, or herbivores (plant-eaters). These animals are then eaten by carnivores (meat-eaters), which are called secondary consumers. Some members of the web eat both plant and animal foods and are called omnivores.

These whales are filterfeeders. They sift out the food from the water through a baleen, which is a fringe of flexible plates around the mouth.

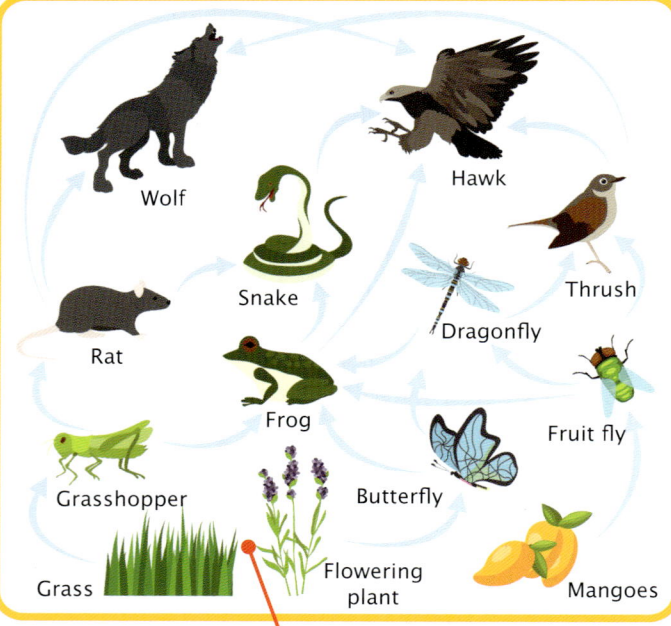

The top of the food web is occupied by the "apex predators," such as wolves and birds of prey.

A pod of Bryde's whales are taking gulps of seawater filled with shoals of small fish. The whales are secondary consumers in the ocean food web.

DID YOU KNOW? A food web is also a map of how energy flows through an ecosystem keeping the wildlife alive. Almost all that energy originally comes from the Sun.

An ocean food web is based on phytoplankton, which are tiny, photosynthesizing organisms that float in the water.

Detritivores

Plants rely on a supply of nonliving chemical nutrients in the soil. These are put there by a group of consumers called detritivores, or "waste eaters." Detritivores eat the waste and remains of other living things, turning them back into soil. Fungi are common detritivores, as are bacteria and flies.

A fungus grows on damp, dead wood, slowly digesting it.

HALL OF FAME:
Rachel Carson
1907–1964

Rachel Carson was an American writer and naturalist who introduced the public to environmental problems caused by pollution and habitat destruction. In 1962, she wrote a book called Silent Spring, where she warned that the chemicals used on the world's farmland were killing so much wildlife that whole ecosystems would collapse. The sounds of animals, like birdsong in spring, would disappear. Thanks to Carson, governments and scientists began to work harder to protect the environment.

Carbon Cycle

All life is based on carbon chemicals. Fat, sugars, proteins, and vitamins are all made up of complex chains of carbon atoms. All organisms take in and give out carbon all the time, and this creates a flow of carbon through the natural environment. This is called the carbon cycle.

Natural Cycle

Carbon is drawn into the food web by plants and other producers that take carbon dioxide gas out of the air or water and turn it into sugars using photosynthesis. This is where all the carbon chemicals in almost every living thing on Earth come from. Life returns some of this carbon dioxide to the environment through respiration. Some of the carbon in the remains of dead organisms becomes locked away in carbon sinks underground. Sometimes, these carbon sinks form coal and oil.

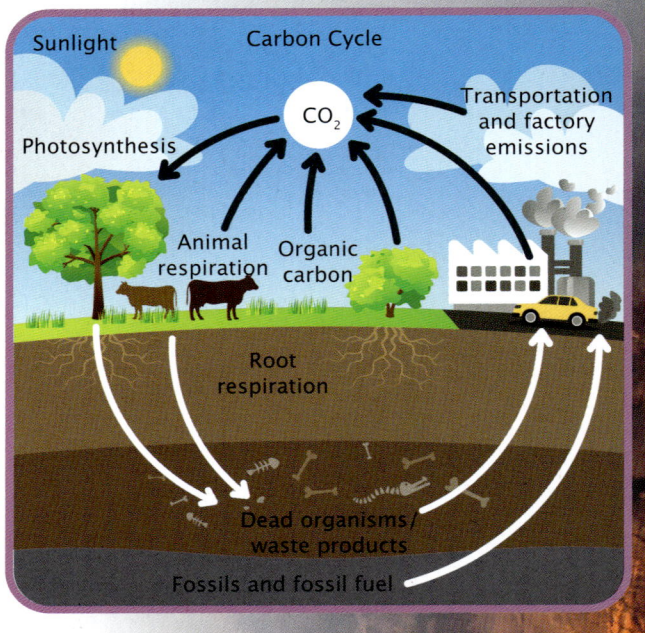

Disrupted Cycle

In the natural carbon cycle, the amount taken in by life is equal to the amount given back out. However, by burning the coal, gas, and oil stored in carbon sinks—known as fossil fuels—extra carbon is added to the air, disrupting the natural cycle. The extra carbon gases in the air traps heat around the planet, making the world warmer and changing the climate.

Climate change is melting the world's ice and glaciers. Life adapted to live in these cold places are running out of habitats.

DID YOU KNOW? The amount of carbon dioxide in the air today is 50 percent higher than it was in the year 1750—and it is still rising.

Climate change is making some regions warmer and drier. Forests become so dry that they catch fire in devastating wildfires.

It will take many decades for the forests to regrow. It may be that the area is now too dry for trees, so grasslands or deserts will replace the burned forest.

Forest wood is an important store of carbon. Wildfires burn the wood and add more carbon dioxide into the air. This makes climate change even worse.

HALL OF FAME:
Eunice Newton Foote
1819–1888

The climate changes caused by disruption to Earth's carbon cycle were first identified by Eunice Newton Foote, an amateur American scientist. In the 1840s, Foote explored how different gases absorb heat from sunlight and found that carbon dioxide got the hottest. She wondered what would happen to the climate if the amount of carbon dioxide in the air went up or down. Foote's work was originally ignored because she was a woman, but her findings were rediscovered in the 1970s.

Other Cycles

Living things need a supply of several chemicals to stay alive. For example, phosphorus is used in DNA and fats, and nitrogen is an essential ingredient in proteins. Plants take these nutrients from the soil, and they are passed on to animals through the food web.

Nitrogen Cycle

The air is 79 percent nitrogen, but this gas is very unreactive and hard for life to take in. The nitrogen cycle relies on bacteria to take the gas from the air and turn it into soil chemicals. Lightning also converts nitrogen gas into chemicals that dissolve in rain. Animal urine and droppings also add nitrogen to the soil. Other bacteria then reverse the process, breaking down nitrogen-rich chemicals in the soil back into pure nitrogen gas.

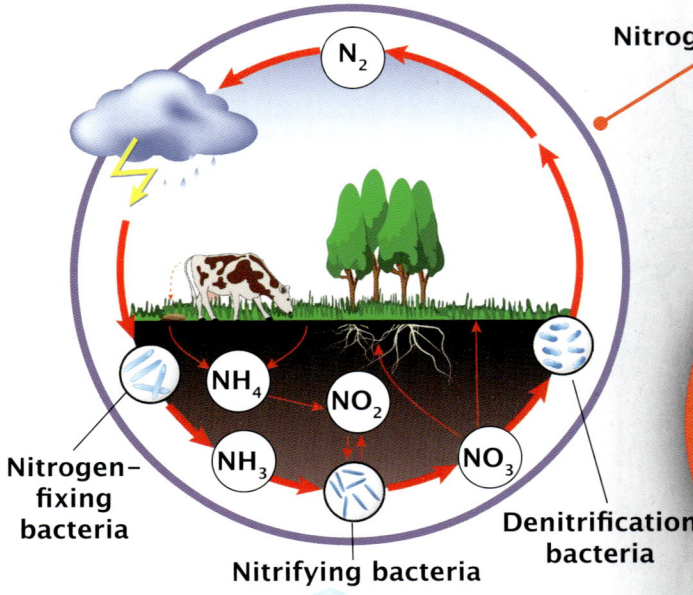

Nitrogen cycle

Nitrogen-fixing bacteria

Nitrifying bacteria

Denitrification bacteria

Most of the land on Earth is higher than the oceans. Gravity causes water to flow into streams and rivers, and then down to the sea.

HALL OF FAME: James Lovelock 1919–2022

James Lovelock worked as a scientist and engineer for nearly 80 years. He is most remembered for his Gaia hypothesis, which is a way of understanding how the entire Earth works as a self-controlling system. Just like a living body, Lovelock described how Earth's systems maintained stable conditions—only these processes took many thousands or millions of years.

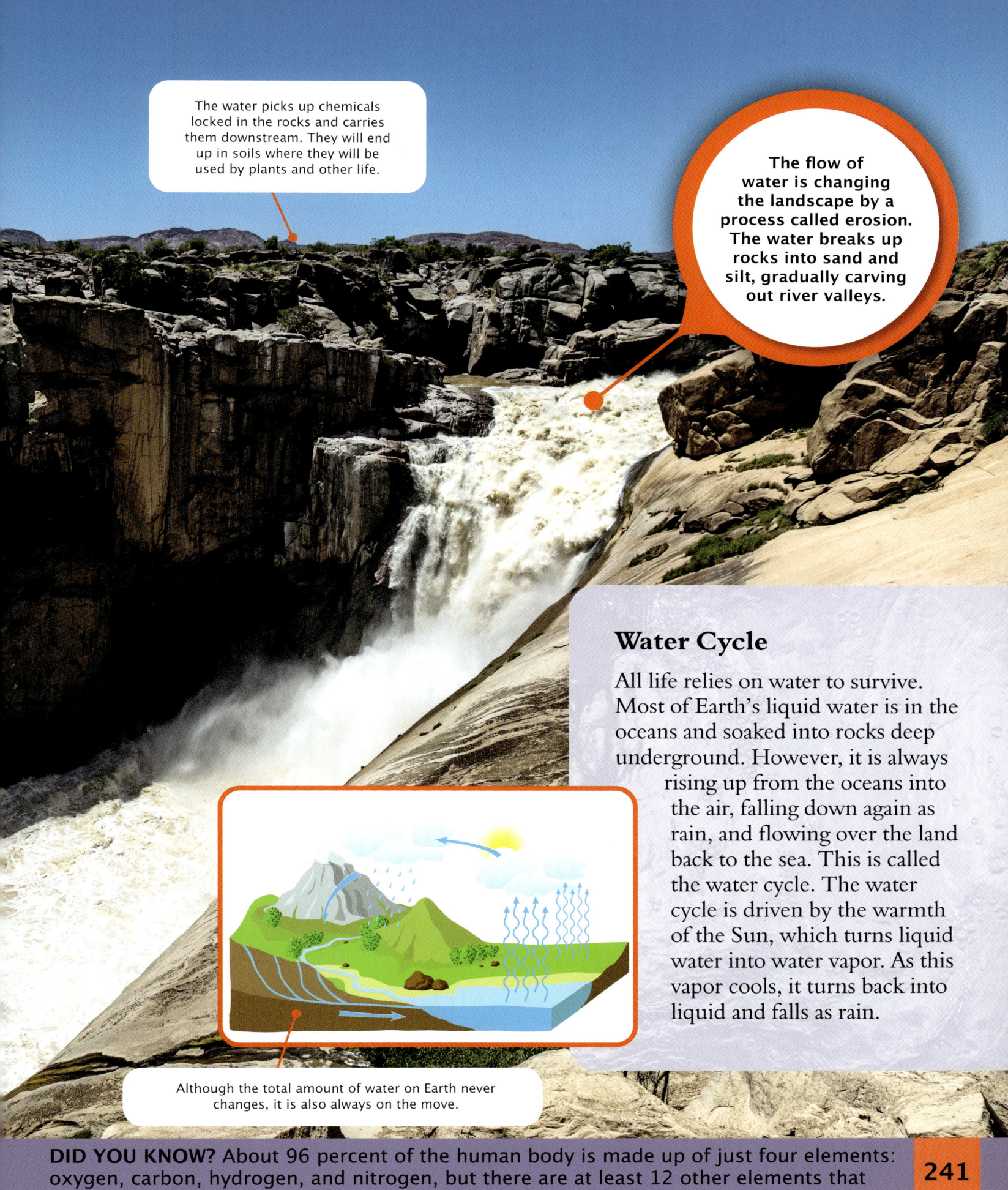

The water picks up chemicals locked in the rocks and carries them downstream. They will end up in soils where they will be used by plants and other life.

The flow of water is changing the landscape by a process called erosion. The water breaks up rocks into sand and silt, gradually carving out river valleys.

Water Cycle

All life relies on water to survive. Most of Earth's liquid water is in the oceans and soaked into rocks deep underground. However, it is always rising up from the oceans into the air, falling down again as rain, and flowing over the land back to the sea. This is called the water cycle. The water cycle is driven by the warmth of the Sun, which turns liquid water into water vapor. As this vapor cools, it turns back into liquid and falls as rain.

Although the total amount of water on Earth never changes, it is also always on the move.

DID YOU KNOW? About 96 percent of the human body is made up of just four elements: oxygen, carbon, hydrogen, and nitrogen, but there are at least 12 other elements that are essential for life.

Dry Biomes

There is a close relationship between a place's climate and the kinds of wildlife communities that survive there. The world is divided into sectors known as biomes based on this. Dry biomes, where there is not much rainfall each year, include deserts, grasslands, and also the polar ice sheets.

Camels are famous desert animals. The hump is filled with oily fats that can be used to provide the camel with food and water for several days.

Low Rain

A desert is any area where less than 25 cm (10 in) of rain falls each year. Grasslands and semi-deserts get more rain but not enough for trees and forests to grow. Low rain is sometimes caused by rain shadows, where all the moisture in the air falls as rain on one side of tall mountains. Only dry air reaches the other side. Dry lands are also found far inland where no rainclouds can reach. Large, hot deserts form in the regions either side of the Equator.

Bison live on the North American prairies, which are formed by the rain shadow of the Rocky Mountains.

Emperor penguins are the only animals to spend the winter in Antarctica. The males look after chicks at this time, while the females feed at sea.

Cold Deserts

The driest place on Earth is not the Sahara or another hot desert. Instead it is Antarctica, where it is almost always below freezing. That means all the water here is frozen and there is no liquid water at all. Very few plants and animals survive in the Antarctic desert because of the cold temperatures as well as the lack of water. Mostly, the animals here live in the ocean and only come on land to rest after feeding out at sea. They get their water from their food.

HALL OF FAME:
Vandana Shiva
1952–

Vandana Shiva is an Indian environmentalist campaigning for farmers across the world to return to traditional methods, especially in places where water and nutrients are scarce. Shiva suggests that this will boost harvests and help keep soils fertile. However, not everyone agrees with this approach and some say that chemical fertilizers and genetically modified crops will be a better way of feeding the world.

Dry biomes have few plants, and the soil lacks nutrients. Instead it is loose sand, and any rain that does fall will trickle right through.

Camels are well adapted to a life in the dry, sandy deserts of Africa and Asia. Their wide feet spread out, so they do not sink in the sand.

DID YOU KNOW? About 40 percent of the Earth's land is covered in dry biomes, like deserts and prairies, which receive less than 130 cm (51 in) of rain each year.

Wet Biomes

Land areas with high rainfall develop into either forests or wetlands. The water allows trees—large but slow-growing plants—to take over the area. If the tops of the trees form a single leafy layer, or canopy, the habitat is a forest. In woodlands, there are gaps between the trees. A wetland occurs where the rainwater cannot drain away, and so marsh or swamp forms.

> Rain forest and jungles grow in tropical areas close to the equator. These areas are always warm and wet, so plants can keep growing all year round.

Boreal Forest

The world's largest forests are in cool parts of the world near the poles. Nearly all of this biome is in the northern hemisphere, and so it is called "boreal" forest (boreal means northern). For most of the year, these forests are covered in snow. The trees here are evergreen conifers, which only grow during the short summers.

The moose, the world's largest deer, lives in the boreal forest. Most big animals live in the sea or on savannas and grasslands.

HALL OF FAME: Wangari Maathai 1940–2011

Wangari Maathai grew up in the mountains of Kenya. She set up the Green Belt Movement, which helped the people living in rural Africa—especially the women—to plant more trees and create a more fertile habitat. This tree planting restored forest areas that had been cleared for fields and created new opportunities for local people.

DID YOU KNOW? A fifth of the world's forests have been cut down in the last 100 years.

Deciduous Forest

Forests that grow in a mild climate are deciduous, which means that the trees drop their leaves before winter and grow them again in spring. This prevents the broad and thin leaves being damaged by freezing winter conditions. The spring and summer are longer and warmer here, so the trees have enough time to regrow leaves each year.

The valuable green chlorophyll is pulled out of the leaves before they are dropped for winter. The leaves change to vibrant shades of red, orange, or yellow.

The rain forests are among the oldest biomes on Earth, often many millions of years old. As a result, each small area of forest has a distinct ecosystem.

Rain forests are packed with life, and there may be 1,000 different animals species—monkeys, insects, spiders, and lizards—living in just one tree.

Ocean Zones

More than two-thirds of the surface of Earth is covered in oceans with an average depth of 3.5 km (2.15 miles). Most of the life in the ocean lives within 200 m (650 ft) of the surface and within 200 km (125 miles) of the coast.

Depth Layers

The conditions for life vary with depth. The top layer is the sunlit or epipelagic zone. This is where there is enough light during the day to photosynthesize. Next is the twilight (mesopelagic) zone, which goes down to 1,000 m (0.62 miles). Many animals stay hidden in this gloom during the day and then hunt at the surface by night. Below this is the midnight (bathypelagic) zone, where it is dark 24 hours a day.

> Sardines eat plankton—floating microscopic organisms. The surface waters are filled with plankton and they are an important food source for many sea creatures.

The diversity of wildlife on the oceanic zones

No plants can live in the deep ocean. Animals that live down at the seabed are described as benthic.

This angler fish has a small glowing lure dangling over its head. Smaller fish that come to investigate will get gobbled up.

In the Deep Sea

The deep, dark oceans are a very empty place with not much to eat. Some animals rely on marine snow, which is the constant supply of fragments of waste and dead bodies that fall down from higher up. Other animals attract prey by using bioluminescence to create their own light.

The fish are all trying to wriggle into the middle of the shoal where they are safest from attack.

Sardines have mirrorlike scales that reflect back the blue of the water. This makes them hard for predators to see.

HALL OF FAME: Sylvia Earle 1935–

An American marine biologist, Sylvia Earle is famous for her work with National Geographic as an explorer and educator. However, before that, Earle was a pioneer of deep-sea exploration, and helped build underwater laboratories for teams of scientists to live and work in for weeks at a time. Today, Earle is one of the Ocean Elders, a team of scientists, environmentalists, and explorers who work to protect the oceans from damage.

DID YOU KNOW? Every year, an estimated 11 million tonnes (12 million tons) of plastic is dumped in the ocean—weighing more than 200,000 blue whales.

Symbiosis

Some organisms have evolved to live in a very close relationship with another species. This is known as symbiosis. There are three types. Mutualism is where both partners benefit from the relationship. In commensalism, one organism benefits, and the other is unaffected. Finally, in parasitism, one species benefits at the cost of the other, its host.

> Corals have symbiotic algae living inside them. As the seas warm, the algae leave corals, making them go white and die, a process called coral bleaching.

Lichens

The crusts that grow on rocks and bark in cold and windy parts of the world are a symbiotic partnership between a fungus and microscopic algae. The algae are inside the fungus, which provides a safe place to live. In return, the algae feeds the fungus with sugars made by photosynthesis. Other kinds of fungi buried in the soil are often in symbiosis with the plants growing at the surface.

The fungus has a hard, crusty body that stops the algae inside from drying out.

Parasite and Host

A parasite cannot live without its host, which provides it with a place to live (for at least a while) and a source of food. Hosts may not be killed by the parasite but are weakened by them. Endoparasites, like tapeworms, live inside the body, often in the digestive system or blood supply. Ectoparasites, such as fleas, live on the outside of the body.

A bite from a bloodsucking mosquito might deliver a microscopic parasite that causes the disease malaria.

HALL OF FAME: Meredith Blackwell
1940–

The American researcher Meredith Blackwell is a leading expert on parasitic fungi, such as *Cordyceps*, that take over the bodies of insects. The fungus grows through the body of its host and eventually kills it. Blackwell uses an electron microscope to study this and other interesting symbiotic and parasitic fungi.

A giant clam cannot get all the food it needs by filtering seawater. It has algae living in its soft body, which contribute sugars in exchange for a place to live.

Giant clams always live in clear, shallow waters, so there is plenty of light for its symbiotic algae to photosynthesize.

DID YOU KNOW? American badgers and coyotes are symbiotic. The coyote is able to sniff out the burrow of a burrowing gopher, and the badger's job is to dig it up.

Social Groups

Some animals will spend most of their time by themselves, avoiding other member of their species. However, it is common for other animals to gather into groups. There are many reasons for crowding together, because living in groups has different benefits.

Chimpanzees live in troops made up of a mixture of relatives and friends. The apes are always arguing about who is in charge.

Herds and Flocks

Large animal groups provide safety in numbers. Seabirds gather on cliffs to breed, hoofed animals form vast herds, and fish school in tight shoals all for the same reason. Predators that attack often have difficulty singling out one individual. A solitary animal would be at greater risk. The animals are also all on the lookout for danger and will warn the others if a predator is nearby.

Seabirds like these gannets compete to nest in the middle of the colony, where it is safest.

Eusocial Animals

Ants, bees, and wasps are examples of eusocial animals, where a large colony of related animals work together to raise the young produced by just one member of the group. This highly organized social system is very good at keeping the colony alive during periods of drought or when food runs low. Other eusocial species include termites, which eat wood, and naked mole rats, which eat roots.

Leaf-cutter ants do not eat leaves but feed them to a fungus garden in the nest, and then the ants eat the fungus.

DID YOU KNOW? Locusts form very large groups called swarms. One of the largest swarms ever recorded was in 1954 and had around 10 billion insects.

Chimps communicate with calls and expressions. This face means that the chimp is unhappy about something.

The apes spend a lot of time grooming each other by cleaning away dirt from their fur. This helps to form trusting relationships.

HALL OF FAME:
Nikolaas Tinbergen
1907–1988

The Dutch zoologist Nikolaas Tinbergen was one of the first scientists to study the ways in which animals behave, a field that is now called ethology. Tinbergen wanted to understand why animals acted like they did, especially those that lived in groups. Tinbergen won the Nobel prize in 1973 for his work in revealing why social groups worked.

Glossary

ACCELERATION
How quickly an object changes its speed, either getting faster or slowing down, and its direction.

ACID
A chemical with a value lower than 7 on the pH scale.

ALKALI
A solution with a value higher than 7 on the pH scale. Alkalis neutralize acids, producing a chemical salt.

AMPLITUDE
The height (or depth) of a wave, measured from the middle.

ARTERY
One of the main vessels carrying blood from the heart to the rest of the body.

ATOM
The smallest possible particle of a chemical element.

BACTERIA
A group of single-cell microorganisms, some of which cause diseases.

BIOME
A large community of life suited to a particular climate and landscape.

CARBOHYDRATE
A substance containing carbon, hydrogen, and oxygen, such as a sugar.

CELL
The basic unit of plants, animals, fungi, and microorganisms. Each cell has a nucleus and is surrounded by a membrane.

CELLULOSE
A substance that is the chief part of the cell walls of plants and is used in making products such as paper.

CHEMICAL BOND
A force that holds atoms together. Chemical bonds are made by sharing electrons (covalent bond) or by losing or gaining electrons (ionic bond).

CHEMICAL REACTION
A process in which atoms are rearranged, changing one or more substances into different substances.

CHLOROPHYLL
A chemical that green plants use to help make their food.

CLASSIFICATION
The arrangement of organisms into groups based on their similarities.

CLIMATE
The usual weather for an area over a long period of time.

CLIMATE CHANGE
A gradual change in Earth's average global temperature. Scientists believe that human activity is causing rapid global warming.

COMPOUND
A pure chemical made from the atoms of more than one element.

CONDUCTOR
A material that lets heat or electricity pass through it.

DENSITY
The space a substance takes up (its volume) in relation to the amount of matter in the substance (its mass).

DIFFRACTION
When a wave spreads out after it travels through a gap.

DIGESTION
The process of breaking down food in the body to release essential nutrients.

DISSOLVE
When a solid is mixed with a liquid and it seems to disappear, it has dissolved.

DNA
Deoxyribonucleic acid, a long molecule found in the cells that carries instructions for the structure and function of living things.

DRAG
Another name for air resistance, or water resistance, experienced by an object moving through a liquid or gas.

ECOLOGY
How organisms relate to one another in their surroundings.

ECOSYSTEM
The community of interacting non-living things and organisms in a habitat.

ELECTRICITY
The effects caused by the presence and movement of electric charges.

ELECTROMAGNETISM
The force that works between charged objects. Opposite charges attract each other, while like ones repel each other. Electricity and magnetism are two aspects of electromagnetism.

ELECTRON
One of the three main particles that make up an atom. It has a negative charge.

ELEMENT
A chemical made of a single type of atom. Elements are the simplest chemicals.

ENDOSKELETON
An internal skeleton.

ENERGY
The ability to do work that can be stored and transferred in different ways.

ENGINEER
A person who designs and builds machines and structures.

ENZYME
A chemical that speeds up or slows down the way in which substances react with one another.

EUKARYOTE
An organism that has cells with a nucleus and other separate structures surrounded by membranes.

EVAPORATE
To turn from liquid into vapor.

FORCE
A push or a pull that can change the movement or shape of an object.

FORMULA
The way scientists write down symbols to show the number and type of atoms present in a molecule.

FREQUENCY
The number of waves that pass a point every second.

FRICTION
A force that resists the movement of objects that are in contact with each other.

GENE
A combination of chemicals that carries information about how an organism will appear and behave.

GENERATOR
A machine that can convert kinetic energy into electricity.

GRAVITY
A force of attraction between all objects that have mass.

HABITAT
The natural home environment of a plant, animal, or other living thing.

HORMONE
A chemical that controls bodily processes such as growth.

IMMUNE SYSTEM
The network of organs, chemicals, and special cells that protects the body from disease.

INSULATOR
A material that heat or electricity cannot pass through.

INVERTEBRATE
An animal without a backbone.

ION
An atom that carries an electric charge because it has lost or gained an electron. A cation is a positive ion, and an anion is a negative ion.

ISOTOPES
Forms of an element where the atoms have the same number of electrons and protons, but different numbers of neutrons.

KINETIC
Relating to motion.

MAGNETISM
The property of some materials, such as iron, to attract or repel similar materials.

MASS
The amount of matter in an object.

MATERIAL
What a substance is made of, for example, ceramic, metal, or plastic.

MATTER
Something that has mass and takes up space.

MEMBRANE
A thin, flexible layer of tissue around organs or cells.

METABOLISM
The chemical processes that the body's cells use to produce energy from food, get rid of waste, and heal themselves.

MOLECULE
A group of two or more atoms that are chemically bonded.

MOMENTUM
The tendency of an object to keep on moving. It is calculated by multiplying the mass of the object by its velocity.

MONOMER
A small molecule that links up with others like it to form a larger molecule called a polymer.

NANOMATERIAL
A material no more than 100 nm (0.0001 mm) long or wide. Particles this small are nanoparticles.

NERVE
The part of the nervous system that carries signals.

NEUTRON
A particle found in the nucleus of an atom. Neutrons have no charge (neither positive nor negative).

NUCLEAR FISSION
A process in which the nucleus of an atom splits into two smaller nuclei, releasing energy as it does so.

NUCLEAR FUSION
A reaction when two nuclei join together to form a single, bigger nucleus, releasing energy in the process.

NUCLEUS
The central part of an atom, made up of protons and neutrons.

NUTRIENTS
Substances that provide food needed for life and growth.

OPTICS
The study of how rays of light behave.

ORBIT
The path of a body around a more massive body, such as that of the Moon around the Earth or the Earth around the Sun.

ORBITAL
An energy shell around the nucleus of an atom, where electrons move around in a wave.

ORGAN
A group of tissues that work together to do a specific job, such as the heart or brain.

ORGANELLE
A part of a cell that does one job.

ORGANIC CHEMICALS
Carbon-based compounds. Living bodies are built from organic chemicals.

ORGANISM
A living thing, including plants, animals, fungi, and single-celled life forms.

OXIDATION
A reaction in which a chemical gains oxygen atoms. The substance that gains oxygen is said to be oxidized.

pH
Almost all liquids are either acids or alkalis, on a pH scale of 0–14. From 0–7 are acids; from 7–14 are alkalis. Neutral substances such as pure water are 7 on the scale.

PHOTON
A particle of light that transmits electromagnetic force from one place to another. A photon has no mass or electric charge.

PHOTOSYNTHESIS
The process of plants using sunlight to create sugars out of water and carbon dioxide.

PITCH
A measure of how high or low a sound is.

POLLINATION
The transfer of pollen so that plants can reproduce.

POLYMER
A very large, chain-like molecule made of repeated smaller molecules.

POWER
A measure of how fast work is being done and energy being used.

PRESSURE
The amount of force acting over an area.

PROKARYOTE
A single-celled organism with no distinct nucleus or cell membrane, such as bacteria or archaea.

PROTEIN
One of the most important of all molecules in the body and in nature. Protein is needed to strengthen and replace tissue in the body.

PROTON
A particle with a positive charge, found in an atom's nucleus.

RADIATION
An electromagnetic wave or a stream of particles that comes from a radioactive source.

REACTIVITY
A measure of how easily a substance reacts with other substances.

REFLECTION
The change of direction of a wave when it bounces off a barrier.

REFRACTION
The change in direction of a wave when it moves from one material to another, such as when light moves from air into water.

RELATIVITY
A set of ideas in physics that explains how space and time are linked. Energy moving through space makes it bend into a different shape and time to change its speed.

RENEWABLE ENERGY
Energy from sources that never run out, such as solar and wind power.

REPEL
To make something move away.

RESISTANCE
A measure of how much a substance blocks, or resists, the flow of electricity.

RESPIRATION
Breathing and also the metabolic process that releases energy from sugars.

RESPIRATORY SYSTEM
The organs that are involved with breathing.

SEMICONDUCTOR
A material that lets electricity pass through it under some conditions.

SOLUBLE
A material that dissolves in water. Something that won't dissolve is insoluble.

SOLUTION
Created when a substance dissolves in a liquid.

SPECIES
A group of similar-looking organisms that can reproduce together.

SPECTROSCOPE
An instrument used to separate and study the wavelengths of light.

SPERM
A male reproductive cell that combines with a female's egg to produce a baby.

STATES OF MATTER
Solid, liquid, or gas. Matter takes a different state, depending on how its molecules are arranged.

STATIC ELECTRICITY
An electric charge held on an object as the result of a gain or loss of electrons.

SUBATOMIC
Smaller than an atom.

SUBATOMIC PARTICLES
Particles inside an atom, including electrons, protons, and neutrons.

TISSUE
A collection of cells that look the same and have a similar job to do in a body.

VACUUM
A space without any matter in it.

VARIATION
The differences in characteristics between individuals of the same species.

VEIN
One of the main vessels carrying blood from different parts of the body to the heart.

VELOCITY
A measure of an object's speed and direction.

VERTEBRATE
An animal with a backbone.

VIRUS
A tiny organism that reproduces inside living cells, often causing illness.

VITAMIN
A natural substance found in foods that the human body cannot usually produce. Vitamins are necessary for good health.

VOLTAGE
A measure of the force that pushes an electric current through a material.

WAVELENGTH
The distance between one wave crest (or trough) and the next.

WEIGHT
The force due to gravity felt by an object with mass.

WORK
A measure of how much energy is being used.

ZOOLOGY
The study of animals and animal life.

Index

acceleration 118, 124–5
air resistance 90
algal blooms 164, 222
allotropes 23
alloys 51
amoebas 164–5
amphibians 174–5
amplitude 127
animals 158, 186–209
    bodies 186–7
    cells 214–15
anions 28
arachnids 173
archaea 163
arthropods 172–3
atmospheric pressure 11, 92
atomic mass 39
atomic nucleus 18–19, 26, 56–7
atomic number 18, 20, 39
atoms 6–9, 14, 18–20, 22, 24, 26, 28, 30, 32, 34, 38, 40, 44, 47–50, 54, 58, 80, 82–3, 99, 105

bacteria 70–1, 162–3, 216–17, 222, 240
balance, sense of 205
binary fission 222
binomial system 160
biology 7, 158–79, 180–209, 210–33, 234–51
biomes 242–5
birds 176–7, 232–3, 250
blood clotting 206
blood vessels 196–7
boiling 14–15
bones 76, 198–9
brain 202–3
breathing 194, 197

carbon 66–7, 238–9
carbon dating 56
carbon dioxide 66–7, 180, 238–9
cations 28
cell division 222–3, 226–7
cell membrane 212, 214, 217, 229
cell walls 212
cells 56–7, 159, 162, 210–33
ceramics 36
chemical bonds 8, 14, 22, 26–9, 34, 48
chemical equations 30
chemical formulae 24
chemical names 24
chemical reactions 30–5, 41, 105, 220–1
chemistry 6, 8–37, 38–57, 58–77
chimpanzees 250–1
chloroplasts/chlorophyll 180, 212, 245
chromosomes 222–3, 224–6
ciliates 164
circuits 53, 150–1, 156
circulatory system 196–7
classification 158, 160–79

climate change 238–9
combustion 32, 35, 114–15
commensalism 248
compounds 9, 16, 24–5, 29, 47, 54
condensation 14–15, 64
conduction 98–9
conductors 21, 29, 36, 50, 52, 144–5, 157
consumers 236
convection 98
coral reefs 234–5, 248–9
corrosion 33
coughing 195
covalent bonds 26–7, 28
crossing over (recombination) 226
cytoplasm 21, 214, 217–19, 223

dark matter 94–5
deformations, elastic/plastic 102
density 10–11
deoxyribonucleic acid (DNA) 77, 162, 190, 212, 216–17, 222, 224–6, 228–9, 233
deserts 242–3
detritivores 237
diatomic molecules 22
diatoms 164
diffraction 140–1
diffusion 13
digestion 192–3, 201
disease 163, 195, 206–7
displacement 31, 50
dissolving 16–17
distortion 140–1
Doppler effect 140–1
drag 90–1, 116

ear 204–5
Earth 60–1
ecology 234
ecosystems 159, 234–51
effort (force) 108–9, 112
eggs 190, 208, 227, 228–9
electric current 82, 102, 104, 142, 146–7, 150, 152, 156–7
    AC/DC 147
electric power 152–3
electrical charge 18, 28, 52, 82, 102, 142–3, 146
electricity 79, 82, 102, 104, 142–57
    static 142–3, 146
electromagnetic spectrum 82, 130–1
electromagnetism 82–4, 152–3
electronics 156–7
electrons 9, 18–20, 22, 26, 28–9, 38–40, 44, 46, 49–50, 52, 56–7, 82–3, 102, 104, 142, 144, 146, 157
electrostatic forces 28
elements 9, 16, 18–21, 24, 38–58, 68–9
endocytosis 218
endothermic reactions 34–5

energy 78–9, 80–128, 133, 140–2
    conservation of 104
    see also electricity
energy shells 18–19, 26, 28, 38–9, 40, 44, 46
energy transfer 96, 98, 100, 106
engines 114–15, 120
entropy 96
enzymes 192, 218, 220–1
evaporation 14–15, 64
evolution 230–3, 234
excretion 192–3
exocytosis 218
exothermic reactions 34–5
experiments 6
eye 138–9, 204

feathers 176–7, 233
fertilization 227, 228–9
fish 175, 186, 246–7
flagellates 164
flight 188
food webs 236–7
force multipliers 108, 118
forces 28, 78, 83, 88–93, 108–9, 112, 116–25
forests 239, 244–5
fossil fuels 43, 154, 238
fossils 230
freezing 14
frequency 126–7
friction 90–1, 116
fuel 42–3, 154, 238
fungi 74, 168–9, 237, 248

galaxies 59, 94–5
gases 10–15, 18, 48–9, 80
gears 114
generators 153, 154
genes 225, 226, 230–1
genetics 159, 190, 210–33
germination 185
global warming 238–9
glucose 66, 68, 72–3, 180
Golgi apparatus 214, 215, 218
gravity 86–9, 94, 96, 102–3, 116

Haber process 70
habitats 159, 234–51
half-life 56
halogens 46–7
haploid cells 227, 228
hearing 204
heart 196–7
helium 26, 38, 42, 48–9, 58–9
hibernation 235
hormones 74, 218
human biology 76–7, 83, 191–209
hydrocarbons 54
hydroelectricity 154
hydrogen 22, 24, 26–7, 38, 40–3, 54–5, 58–9

immune system 206–7
inertia 117
inorganic chemicals 54–5
insects 76–7, 172
insulators 144–5, 149
interconnection 236
interference 132–3
invertebrates 76, 170–1
ionic bonds 28–9
ions 28–9, 144
isotopes 19, 39, 56, 58

joints 198

kinetic energy 96, 98, 100–1

leaves 182
LEDs 156–7
lenses 138–9
levers 108–9
light 82, 105, 129–41
lightning 148–9, 240
liquids 10–12, 14–15, 18, 20, 80
load 108
lock and key theory 220
locomotion 188–9

machines 108–13, 114–15, 120
magnetic fields 84–5, 152
magnetism 82, 84–5, 152
magnifying lenses 138–9
mammals 178–9
mass 10, 86, 88–9, 118–19, 122
mass number 19, 39
matter 6, 8, 78, 80–95
states of 10–37, 80
meiosis 226–7
melting 14–15
menstruation 208
metals 29, 36, 50–1, 98–9, 144
alkali 40–1
alkaline earth 44–5
post-transition 50
separating 50
transition 50
microchips 53, 84
microscopes 210–11
minerals 62–3
mitochondria 180, 214
mitosis 222
mixtures 16–17
molecules 8, 14, 22–4, 30, 54
mollusks 170–1
momentum 122–3
monomers 55
monosaccharides 72
motion 78, 83, 96–125, 152, 188–9
laws of 116–21
motors, electric 152
muscles 83, 200–1
mutualism 248
mycelium 169

natural selection 230–2
neon 48–9
nervous system 202–3

neutrons 9, 18–20, 38–9, 56–7
nitrogen 70–1, 240
noble gases 48–9
noise-cancelling technology 133
non-metals 52–3
nuclear fission 57
nuclear fusion 58

oceans 246–9
omnivores 236
optics 79, 126–41
organelles 214–15, 222, 229
organic chemicals 54–5
oscillation 129
osmosis 218–19
oxidation 33, 41
oxygen 66–9, 180–1, 194

parasitism 248
particle theory 12–13
pathogens 206–7
Periodic Table 9, 38–57
peristalsis 201
photosynthesis 66–9, 72, 74, 164, 166–7, 180–1, 235, 249
physics 7, 78–95, 96–125, 126–41, 142–57
planetary orbits 86
plants 66, 158, 166–7, 180–5, 218
bodies 182–3
cells 212–13
chemicals 74–5
plasma 80, 146–7
plastics 36–7, 55, 144–5
poisons 47, 74–5, 168
pollination 184–5
polymers 8, 36, 55, 76, 212, 220, 225
potential energy 96, 102–3, 105
power 106–7, 152–5
pregnancy 209
pressure 11, 92–3, 104
properties 10, 24, 36–7
proteins 76
protists 164–5, 212
protons 9, 18–20, 38–40, 56–8
pulleys 112

radar 134
radiation 82, 98
radio waves 130, 134
radioactivity 56–7
ramps (inclined planes) 110–11
reactivity 31, 41–2, 44–6, 50
reflection 133, 134–6, 140
reflexes 202
refraction 136–7
renewable power 154–5
reproduction 184–5, 190–1, 208–9, 226–9
reptiles 174–5
resistance 89, 90–1
respiration 68–9, 180–1, 238
respiratory system 194–5
rivers 64–5, 154
rocks 62–3, 241
roots 182–3
rubber 36
rusting 33, 34

salt 29
saprophytes 168
scientists 6–7
screws 110–11
sea anemones 187
semi-metals 52–3
semiconductors 52, 156–7
senses 202, 204–5
sex cells (gametes) 184, 226, 228–9
sexual dimorphism 232
sexual selection 232–3
silicon 52–3, 157
single-celled organisms 162–5
skeletons 76, 198–9
smell, sense of 205
social groups 250–1
solar power 154–5
solids 10–12, 14–15, 18, 20, 80
solutions 16–17, 17
solvents 17, 219
sound energy 104, 127–8, 133, 140–1
space 89
specialization 214
species 158, 160–1
speed 124–5, 127
sperm 208, 227, 228–9
stars 58–9, 86, 140
surface tension 65
symbiosis 248–9

taste, sense of 205
taxonomy 160–1
temperature 14, 98
thermal decomposition 32, 35
thermal (heat) energy 96, 98–9, 105
touch, sense of 204
transformers 148
transistors 53, 156–7
turbines 153, 154–5

ultraviolet (UV) 130

valency (bonding ability) 26, 48
velocity 100–1, 122, 124–5
terminal 87
vertebrates 76, 174–9
viscosity 91
vision 202, 204
voltage 148–9
volume 10

water 64–5, 92, 128, 154, 182, 219
water cycle 64, 240–1
watts (W) 106–7
wavelength 126–7, 129, 130–1, 133, 136, 140
waves 79, 104, 126–41
wedges 110
weight 88–9
wheels 112–13
white blood cells 206
wind turbines 154–5
wires 144–5
Wood Wide Web 74
work, doing 96–125